Wetlands and Western Cultures

Environment and Society

Series Editor: Douglas Vakoch

As scholars examine the environmental challenges facing humanity, they increasingly recognize that solutions require a focus on the human causes and consequences of these threats, and not merely a focus on the scientific and technical issues. To meet this need, the Environment and Society series explores a broad range of topics in environmental studies from the perspectives of the social sciences and humanities. Books in this series help the reader understand contemporary environmental concerns, while offering concrete steps to address these problems.

Books in this series include both monographs and edited volumes that are grounded in the realities of ecological issues identified by the natural sciences. Our authors and contributors come from disciplines including but not limited to anthropology, architecture, area studies, communication studies, economics, ethics, gender studies, geography, history, law, pedagogy, philosophy, political science, psychology, religious studies, sociology, and theology. To foster a constructive dialogue between these researchers and environmental scientists, the Environment and Society series publishes work that is relevant to those engaged in environmental studies, while also being of interest to scholars from the author's primary discipline.

Recent Titles in the series

Wetlands and Western Cultures: Denigration to Conservation, by Rod Giblett
Sustainable Engineering for Life Tomorrow, edited by Jacqueline A. Stagner and David S. K. Ting
Nuclear Weapons and the Environment: An Ecological Case for Nonproliferation, by John Perry
Portland's Good Life: Sustainability and Hope in an American City, by R. Bruce Stephenson
Conservation, Sustainability, and Environmental Justice in India, edited by Alok Gupta
Environmental and Animal Abuse Denial: Averting Our Gaze, edited by Tomaž Grušovnik, Reingard Spannring, and Karen Lykke Syse
Living Deep Ecology: A Bioregional Journey by Bill Devall, edited with an introduction by Sing C. Chew
Secular Discourse on Sin in the Anthropocene: What's Wrong With the World? by Ernst M. Conradie
Australian Wetland Cultures: Swamps and the Environmental Crisis, edited by John Charles Ryan and Li Chen

Wetlands and Western Cultures

Denigration to Conservation

Rod Giblett

LEXINGTON BOOKS
Lanham • Boulder • New York • London

Published by Lexington Books
An imprint of The Rowman & Littlefield Publishing Group, Inc.
4501 Forbes Boulevard, Suite 200, Lanham, Maryland 20706
www.rowman.com

6 Tinworth Street, London SE11 5AL, United Kingdom

British Library Cataloguing in Publication Information Available

Library of Congress Cataloging-in-Publication Data Available

ISBN 978-1-7936-4345-2 (cloth)
ISBN 978-1-7936-4346-9 (electronic)

Contents

Preface

Wetlands and cultures—wetland cultures, the cultures of wetlands—including western cultures and traditional Australian Aboriginal and Chinese Taoist cultures, have been a longstanding interest of mine for three decades beginning in 1990. This interest was generated and pursued in both my conservation and environmental activism and my academic research and writing. My early work on wetlands was published in both academic journals (Giblett 1992, 1993a and b) and conservation magazines (collected in Giblett 2013c). The introduction to *Western Australian Wetlands* (Giblett and Webb 1996) was first published in the conservation press (Giblett and Webb 1993). Similarly, the short discussion of Forrestdale Lake in *Western Australian Wetlands* (Giblett and Webb, eds, 1996, 109, 113–114, 116) was first published in the conservation press (Giblett 1993c).

Writing and publishing academic and activist articles and books, such as *Postmodern Wetlands* (Giblett 1996), at about the same time as each other seemed important to me at the time to try to cross the great divides between academia and activism, and between the sciences and the humanities, and to make the link between them with the aim that activism might be better informed critically, culturally, historically and philosophically, and that academic considerations of the history and culture of wetlands might flow through into action in conservation, protection, and rehabilitation (see Giblett 2013c).

Postmodern Wetlands is a pioneering study in the environmental humanities and the first book of wetland cultural studies and psychoanalytic ecology claims substantiated in the following introduction. It is also transdisciplinary across literature, philosophy, history, and ecology (as is this book). In addition, it engages with Freudian psychoanalysis (as does in this book), as well as French feminism and post-structuralism. Its geographical areas of interest and inquiry were principally English, American and Australian (as is this book too with different topics and case studies). It also engaged with the cities of London and Perth (Western Australia), my home city at the time. It is now regarded as a classic in the environmental humanities. This may mean that no one reads it, but everyone who has heard of it thinks they know what it is about!

For twenty-eight years, my life and work at and by Forrestdale Lake, an internationally important wetland in Western Australia, helped me to cross the academic/activist and humanities/sciences divides. This was achieved mainly by co-founding the Friends of Forrestdale with David James in 1990, co-writing an article with him many years later and being its Secretary for twenty-five years (see Giblett 2006, 2013a; Giblett and James 2009). Starting from 1990, my academic and activist life and work went hand in hand (see Giblett 2013c). The beginnings of my recent work, perhaps regrettably, have not been in environmental activism, nor in the conservation and environmental press for various personal and biographical reasons.

My recent work has still had a bioregional focus on the city of Melbourne in which I now live. This focus resulted in initial local publication in an article about Melbourne's lost and found wetlands (Giblett 2016b) and in the first part of a recent book about Melbourne as a "City of Ghost Swamps" (Giblett 2020c, chapters 2–5). This wide-ranging environmental and cultural history and study of Melbourne also situates it in its bioregion of the Yarra River catchment (Giblett 2020c, chapter 8) and considers its built and cultural environment and history, including its arcades, boulevards, parks, gardens, trams, trains, sporting obsession and venues, and exhibitions and exhibition buildings (Giblett 2020c, chapters 6, 7, and 9–11).

Postmodern Wetlands and *Black Swan Lake* included a case study of the city of Perth, the capital city of Western Australia, revising and updating previous work (Giblett 1993a, 1996, chapter 3; 2013, chapter 15). *Western Australian Wetlands* also included a case study of the city of Perth in a different, less academic, more locally focused version (Giblett and Webb, eds, 1996, 127–146). Colleagues and collaborators have recently followed in these footsteps in their studies of Perth as a city of wetlands (Brady and Murray, 2020; Ryan, Brady and Kueh, 2020).

My discussion of wetlands in the western third of a continent in one state of Australia and its capital city grew into an interest in the wetlands through the lower provinces of Canada from the Pacific to Atlantic coasts of another continent resulting in research for, writing and publication of *Canadian Wetlands* (Giblett 2014). The discussion in *Postmodern Wetlands*, *Black Swan Lake*, and *Western Australian Wetlands* of Perth as a city of wetlands in the past and present also blossomed into a longer and deeper consideration of other swamp cities and marsh metropolises, such as Toronto in an article and in *Canadian Wetlands* (Giblett 2013b, 2014, chapter 7; 2016a, chapter 12) and many cities in Europe (Paris, London, Venice, Berlin, Hamburg, and St Petersburg) and the United States (Boston, New York, New Orleans, Chicago and Washington) in *Cities and Wetlands* (Giblett 2016a).

Wetlands had become so pervasive in my work in the 1990s that I was typecast as "the swamp man." For several years from the twenty noughties I pursued other interests in the human body, especially tai chi (Giblett 2008a and b), water, especially its cross-cultural color-coding (Giblett 2007, 2009, chapter 11), technology, especially photography (Giblett 2008c, 2009, 2011; Giblett and Tolonen 2012), and ecocultural studies (Giblett 2012), only to return to wetlands later in the twenty teens (Giblett 2013a–c, 2014, 2016a and b, 2017).

Wetlands were then a departure point for the exploration of other topics in a variety of genres, such as dragons in tales and a saga as creatures of the wetland in the form of either marsh monsters or swamp serpents (Giblett 2018a, c; 2019a, c), the legend of St George and the dragon (Giblett 2018a–c; 2019a–c), environmental theology (Giblett 2018b) and the uncanny (Giblett 2019b, d), only to return full circle to wetlands more recently in writing about wetlands in the environmental humanities (Giblett 2020a, b; and the present volume), including consideration of wetlands in Australian Aboriginal wetland cultures and stories that had its beginnings in the 1990s (Giblett 1996, 2020e). Wetlands also figure in my recent work on conservation and the environmental apostles (Giblett 2020d) and in a collection of detective stories entitled *Swamp Deaths: Collected Cold Cases and Other Marshy Mysteries* (Giblett forthcoming) in which I introduce Inspector Colin Thorow and his memoirs of old marshy mysteries (in new meta-detective stories).

Returning full circle to wetlands in the present volume is reinforced by the fact that most of the contents of the present volume had their beginnings in *Postmodern Wetlands* (Giblett 1996). Chapter 1 had its beginnings in *Postmodern Wetlands* in a short discussion of *Beowulf* and a scattered discussion of Milton's *Paradise Lost.* In an early draft of *Postmodern Wetlands* I also discussed the Dead Marshes in Tolkien's *Lord of the Rings*, but this was later cut for reasons of length and lost. Writing chapter 1 for the present volume has enabled me to produce a richer reading of *Beowulf, Paradise Lost* and Tolkien's work, and in the case of *Paradise Lost*, an environmentally and wetland friendly rewriting. It also enabled me to place the Dead Marshes in close context with *Beowulf* and my work on the trench landscape of the World War I in *Landscapes of Culture and Nature* (Giblett 2009).

Chapter 2 also had its beginnings, as did psychoanalytic ecology, in *Postmodern Wetlands.* Of course, psychoanalytic ecology had its beginnings before the term was invented (*avant la lettre*) in Sigmund Freud's work on sublimation, symptom, mourning, melancholia, and the uncanny as discussed in *Postmodern Wetlands* and developed in this chapter. Chapter 2 also had its beginnings in *Postmodern Wetlands* in the discussion of Bunyan's "slough of despond" and Burton's "anatomy of melancholy." Chapter 3 had its beginnings in *Postmodern Wetlands* too in its scattered discussions of the Fens and Graham Swift's *Waterland.*

Chapter 4 had its beginnings in a discussion of Eugene von Guérard's painting, "Mount William and part of the Grampians in West Victoria," and Arthur Boyd's painting "Irrigation Lake, Wimmera" in a recent chapter on wetlands in Australian painting and photography (Giblett 2020a). *Postmodern Wetlands* discussed aesthetics, in particular the sublime in the work of Longinus, and the sublime and the beautiful in work of Edmund Burke and Immanuel Kant in relation to the wetlandscape. Chapters 4 and 5 of the present volume build on this general work on aesthetics and apply it to the aesthetics of specific dry and wet landscapes. Chapters 6 and 7 apply work on cities and wetlands begun in *Postmodern Wetlands* and developed since then in other books. Chapter 8 also had its beginnings in a short section of *Postmodern Wetlands* devoted to Thoreau as "the patron saint of swamps." Chapter 9 reflects on my own nature writing in *Black Swan Lake,* based, in part, as Thoreau did, on the nature journal I kept for a couple of years (Giblett 2013a).

Wetlands and Western Culture is a sort of "Postmodern Wetlands Part Two" covering a similar area and engaging with the same disciplines and approaches of cultural studies, philosophy, history, literature, and psychoanalysis. Similarly, it also engages with a variety of geographical areas and with a number of cities. *Wetlands and Western Culture* develops the preliminary discussion of the theology and psychology of wetlands in *Postmodern Wetlands*. It also extends its range into case studies of specific sites and cities. It is a stand-alone volume with an updated theoretical introduction to follow and new cultural discussions in the subsequent chapters. It refers to my previous work where pertinent for the background, further discussion and greater elaboration of the underpinnings of the present volume and its larger context (hence the self-references). It is a celebration of twenty-five years since the publication of *Postmodern Wetlands* and the culmination of thirty years of studying and writing about wetland cultures in the environmental humanities. It is my academic black swan song (see also Giblett 2021).

Coburg, 2020

Acknowledgments

I begin by making a general and grateful acknowledgment of Zoë Sofoulis for her generous sharing of her work on the s(ub)lime and the uncanny, for pointing out the pertinence of Norman O. Brown's work on the sublime city and for many other suggestions acknowledged in previous books. The theoretical framework for the present volume as articulated in chapter 1 and as depicted pictorially in figure 3.1, "A Psychogeocorpography of Modernity," is based on Zoë's work. My work is a footnote to her work. She does not bear any blame or responsibility for what I have done with it though.

For the Introduction I am grateful to Friederike Wiebke for lending me her copy of Andrew Darby's *Flight Lines* that reminded me about the importance of wetlands as wader bird homes. I am also grateful to Robert Thorson for pointing out the pertinence of James C. Scott's *Against the Grain* that contains a dozen pages "for wetlands." The present volume is also "for wetlands," which prompted one reviewer to complain that it is more a work of advocacy than argument, probably the nicest backhanded compliment they could have paid it. John Charles Ryan alerted me to the concept of the paludal, for which I am grateful.

For chapter 1, I am also grateful to John for re-igniting my interest in Tolkien's and *Beowulf*'s take on wetlands and for passing on the reference to *The Silmarillion*. I am grateful to Sandra Giblett for drawing my attention to Diana Gabaldon's *An Echo in the Bone* and her account of "The Great Dismal" Swamp. An earlier and shorter version of this chapter was first published as "Theology of Wetlands: Tolkien and *Beowulf* on Marshes and their Monsters," *Green Letters: Studies in Ecocritcism,* 19 (2) 2015: 132–143, © Association for the Study of Literature and the Environmental (ASLE UK-1), reprinted by permission of Taylor & Francis Ltd, www.tandfonline.com on behalf of Association for the Study of Literature and the Environmental (ASLE UK-1). I am grateful to the publisher for granting this permission.

For chapter 2, I am grateful to Juha Tolonen for pointing out the pertinence of Antoine Picon's "Anxious Landscapes: From the Ruin to the Rust."

For chapter 3, I am grateful to Matthew Gandy for drawing my attention to Vittoria di Palma's *Wasteland* and to Sandra Giblett for drawing my attention to Phillip Pullman's description of the Fens in *His Dark Materials.* I am also grateful to Becky Owen-Fisher for her generous permission to quote from her poetry.

For chapter 4, I am grateful to Carole Mules for sending to me a copy of the map in figure 4.3 and of Rod Bird's report on the wetlands of south-west Victoria. I am also grateful to Carole for the opportunity to write about her artwork on Gariwerd wetlands, for her generous permission to reproduce it and for guided tours of the Gariwerd wetlands on many occasions and discussions about them hence the dedication of this book in part to her too.

For chapter 5, I gratefully acknowledge the assistance of Joanne Francis and John Charles Ryan with this chapter, especially with suggestions for research and for supplying other information. I am also grateful to Joanne for her generous permission to reproduce her artwork and email exchanges and phone conversation about them hence the dedication of this book in part to her too. She does not bear any blame for the readings I have given of them.

For chapter 6, I am grateful to Donna Houston and Emily O'Gorman both of Macquarie University in Sydney for their helpful suggestions of references on Sydney wetlands.

For chapter 7, I am grateful to Philip Jennings for drawing my attention to Nigel Bertram and Catherine Murphy's *In Time with Water: Design Studies of 3 Australian Cities* (University of Western Australia Publishing, 2019).

For chapter 8, I am grateful to John Charles Ryan for driving Sandra and I to Concord and Walden Pond in Massachusetts on the final leg of my Thoreauvian pilgrimage in April 2019. Thanks also to both for their companionship on the Thoreauvian way, and for much longer, before and after on the way of life. Hence the dedication of *New Lives of the Saints* to John and the present volume in part to Sandra. I am also grateful to Robert Thorson for permission to cite his email about Wyman Meadow and Thoreau's house and desk.

Finally, I am grateful to several anonymous reviewers for their helpful comments and incisive critique that have helped immensely to improve cohesion, sharpen the argument, and clarify the structure. All blame for the faults in what follows is, of course, mine.

Coburg, Victoria
2020

Introduction

An Invitation for Wading into Wetlands

For many centuries, even a millennium, mainstream western, principally Anglophone, culture denigrated (literally "blackened"[1]) wetlands as dead black waters—as places of darkness, disease and death; horror, slime and the uncanny; melancholy and the monstrous—and destroyed them. Within western culture a counter tradition affirms wetlands as living black waters—as places of both life and death, light and dark, as biologically rich and fertile, mucky and murky, vital for life on earth—and calls for their conservation. Black waters live!

Referring to western cultures in the present volume includes both the mainstream and counter traditions in "the west," meaning western European, primarily English, culture, with its converging, conflicting, and diverging Greco-Roman philosophical and Judeo-Christian theological roots, its settler-colonial and Anglophone diaspora, principally in North America and Australia, inflected in their "pagan" (or pre-, or non-Christian, "pre-patriarchal" western, or matrifocal and gylanic,[2] cultures), Christian and post-Christian moments and histories. Greek thought and Christian theology converged in mainstream western, principally Anglophone, culture and literature with their syncretization in the seventeenth century CE in John Milton's epic poem *Paradise Lost* and its denigration of wetlands (as discussed in chapter 1 of the present volume). Coincidentally, and not accidently, the large-scale, mechanized drainage of the English Fens began in the seventeenth century CE (as we will see in chapter 3 of the present volume). Denigration led to destruction.

THE ELEMENTS AND THE HUMORS

The denigration of wetlands in mainstream western culture[3] began arguably with the ancient Greek Hippocratic writings on medicine. Wetlands were regarded as the source of disease when miasmatic vapours rise from stagnant pools of water and cause malaria (literally "bad air"; see Giblett 1996, chapter 5). The miasmatic theory of disease prevailed up until the 1890s. It was based on the Greek philosophy of the four elements: earth, air, fire, and water. Wetlands are problematic, indeed aberrant, from this point of view because they mix the elements of air and water in miasma and malaria (as well as air and "fire" or heat in the humid tropics). These elements should remain separate. They should stay in their proper places in the clean, disease-free body of the earth and human beings.

Wetlands have also been seen as bad for the mind in mainstream western culture as they can plunge the mind into melancholia, and even into madness. The association between wetlands and melancholy (or with despondency and despair) also has its roots in the philosophy of the elements and their associated humors (psychosomatic affects) formulated in the Elizabethan "world picture" of the late sixteenth century CE (see Giblett 1996, 160). As wetlands mix the elements, they produce an aberrant "humor," strictly a kind of phlegmatic melancholy as anatomised psychologically by Robert Burton and theologised psychologically by John Bunyan in the seventeenth century CE as "the slough of despond," and secularised psychologically by Charles Dickens in the nineteenth century as "the grim domains of Giant Despair" and by E. M. Cioran in the twentieth century as "the languid mud" of melancholy (as discussed in chapters 2, 3, and 6 of the present volume).

Mainstream western culture was founded in part and still functions on the philosophy of the four elements, or, in more modern terms, the material properties of solid, gas, heat, and liquid. Wetlands mix not only air and water in miasma and malaria, but also earth (or solid) and water (or liquid) in slime. They cross the boundary between land and water, and can even be in transition spatially and temporally between open water to dry land (and vice versa). They are a troubling and unsettling category mediating between earth and water, solid and liquid. Wetlands are slimy. Slime is primary, the first and the best act of creation. In the beginning, God created the world as wetland (*Genesis* 1: 1–2 and as discussed in chapter 1 of the present volume).

Slime mixes solid and liquid, like wetlands do. Slime and other substances result from six combinations and transformations of the four elements (or properties) of earth (or solid), air (or gas), fire (or heat) and water (or liquid):

1. solid, earth + liquid, water = slime[4];
2. solid, earth + gas, air = sublime[5];

3. liquid, water + heat, fire = steam;
4. liquid, water + gas, air = miasma;
5. solid, earth + heat, fire = ore;
6. heat, fire + gas, air = firestorm.

Industrial capitalist technologies manipulated these processes to greater or lesser good. They transformed the elements into these combinations with sometimes devastating consequences, such as the firestorms of firebombing in World War II and bushfires (feral or native, arsonist or lightning strikes) in the age of global heating and climate catastrophes.

Industrial capitalist technologies also harnessed the power of other combinations, such as steam, to manipulate and manufacture others, such as metals from ores. They mined the solidified slime of carboniferous swamps when the world was wetland in the form of coal to power steam engines and generate electricity with both producing emissions of greenhouse gases to inaugurate global heating. Later they sublimated the oily slime of carboniferous swamps into the gas of diesel and petrol to power internal combustion engines with their emissions of greenhouse gases contributing to and increasing global heating. Then they fracked and desublimated natural gas into liquefied petroleum gas. These transformations lie at the heart of modernity for a couple of centuries and have their basis in mainstream western culture for a couple of millennia. They are the key to decoding modern western culture. Wetlands now may make up only 6 percent of the earth's surface now, but wetlands of the past when the earth was dominated by swamp forest and when the world was wetland supply 100 percent of the "fossil fuels" needed to power industrial, carboniferous capitalism for the past 200 years.

The fact that this stage of the swamp world is now largely forgotten, but that the industrial world's use of "fossil fuels" during every minute of every day re-affirms its existence constitutes what Walter Benjamin (1999, 462, 464) called "a dialectical image." In his words, "what has been within a particular epoch is always, simultaneously, what has been from time immemorial," in this instance, the swamp world. "The primal phenomenon of history" of the swamp world is "present by virtue of its very oblivion" in what he calls "the now of recognisability." The fact that the vast majority of the users and beneficiaries of "fossil fuels" are oblivious to the swamp world from where and when they come means that it is present in two senses in time and space by virtue of its very oblivion. Similarly with cities built on or in swamps, the fact that the vast majority of their residents are oblivious to the swamp world from where and when they come means that it is present by virtue of its very oblivion.

The transformation of the slimy swamp into the sublime city built on the foundation of the six transformations listed.[6] In mainstream, modern western culture wetlands have been filled or drained because they pose an obstacle to, and to provide the dryland surface for, agricultural and urban development. By contrast, premodern Anglophone culture in the Fens and other wetlands in England used traditional paludicultural (from the Latin word "palus" meaning "mire" or "marsh") and aquacultural techniques to provide sustenance. Similarly Francophone culture both at home and abroad in Canada used the technology of *aboiteaux* to regulate water flows and produce agricultural output from wetlands (see Giblett 2014).

Humans have cultivated wetlands for over 8,500 years. Around 6,500 BCE the rich alluvium of Mesopotamia between the two rivers of the Tigris and Euphrates was what James C. Scott (2017, 44, 47, 127) calls "a [. . .] wetland paradise" where marshlanders practiced "an exuberant diversity of livelihoods." Drawing here on the work of Jennifer Pournelle, Scott (2017, 47) argues that "the earliest large fixed settlements sprang up in wetlands," probably the beginnings of cities in swamps and metropolises in marshes (see Giblett 2016).

Two thousand years later Australian Aboriginal people built the earliest known small fixed settlements near wetlands in Australia. From 1788 CE the colonising settlers built settlements near wetlands too that grew into cities, such as Melbourne, Sydney and Perth, as wetlands provided fresh water and feed for livestock. These wetlands were later destroyed by filling or by being landscaped into parks. By contrast, the Gunditjmara people of Lake Condah in south eastern Australia not only constructed a settlement by building stone houses and engineered aquaculture by constructing a system of stone channels that trapped fish and eels, but also practiced paludiculture by cultivating edible plants in wetlands and attracting waterbirds that they netted to provide sustenance (rather than merely "subsistence"). These wetlands were traditionally a rich source of animal and plant foods (Gunditjmara with Wettenhall 2010, 7, 13–16, 67). This site of stone houses, engineered traps and cultivated wetlands is called "Budj Bim." It has been dated to 6,700 years old, older than Stonehenge. It is now a World Heritage site (AAP 2019).

Continuing to draw on the work of Pournelle, Scott goes on to relate that the Mesopotamian settlements "relied overwhelmingly on wetland resources [. . .] for their subsistence" (Scott 2017, 47). Or more precisely, they relied overwhelmingly on wetland resources for their sustenance as these wetlands were traditionally, as Scott (2017, 47–57, 127–128) goes on to discuss at length and as they were for the Gunditjmara people at Lake Condah, a rich source of animal and plant foods (as they and Wettenhall discuss at length too). As both peoples regulated the flows of water and cultivated water plants in wetlands, they were not only hunters, gatherers, and foragers (as Scott calls the Mesopotamian marshlanders and as Australian Aboriginal peoples have also been called), but also aquacultural engineers and paludicultural practitioners. The wetlands were also "commons" that both peoples owned and used in common. Scott (2017, 57) refers to the Mesopotamian marshlands in scare quotes as "common property resources."[7]

THE HORROR OF WETLANDS

Besides being seen as places of physical and mental illness, wetlands have often been regarded in mainstream western culture as home to some sort of horrifying marsh monster or swamp serpent lurking in their murky waters as portrayed, for instance, in the Old English *Beowulf* dating from the eleventh century of the CE (as discussed in chapter 1 of the present volume). Monsters are a taxonomic anomaly in a classificatory system, or "the order of things," based on a hard and fast distinction between human and animal, real and imaginary creatures. They cross the boundaries between, and mix the elements of, earth, air, fire, and water. They are evil beings of dead black waters.

By contrast, in indigenous Australian Aboriginal cultures wetlands have been seen (and continue to be seen) as places of both life and death, light and darkness, life-giving and death-dealing to be revered and conserved—in other words, as "living black waters." They are often associated, or equated with, with the usually brightly.colored rainbow, but sometimes black, yet richly patterned serpent or spirit of the Dreaming stories.[8] Australian Aboriginal peoples have also employed agricultural, aquacultural, and paludicultural practices of livelihood for thousands of years. The earthly sacrality of marsh monsters and swamp serpents are figures for sustaining and nourishing a belief in, and a way of living and practicing, bio- and psycho-symbiotic livelihoods in bioregions of the living earth in the Symbiocene, the hoped-for age and paradigm superseding (in the sense of embracing and following) the Anthropocene. The Symbiocene comes out of the desire for, and paradigm of, mutuality. The Anthropocene came out of its drive for, and the failed paradigm of, mastery over nature.[9] In the era of climate change and global warming, nature is mastering, or monstering, we humans. The new, anthropogenic climate regime rules—and it is *not* okay.

Along similar lines to Australian Aboriginal cultures, other non-western, and "pre-patriarchal" western, or matrifocal and gylanic, cultures feminized the swamp positively as the source of new life in the great mother, the snake goddess, the mistress of living black waters, whereas patriarchal hierarchy with its dryland agriculture and its misogynist denigration of the wetland feminized the wetland as the environmental *femme fatale*, spider woman and *vagina dentata* (see Giblett 1996, 145–150). With the rise of capitalism under the aegis of patriarchy in Europe with its modern cities the black waters of wetlands "at home" and in the colonies abroad were seen by many citizens as a pre-modern wasteland or wilderness to be conquered as a marker of "Progress." Wetlands were drained or filled to create the dead surface of private property on which agricultural and urban development could then take place, or they were polluted by cities and farms to produce the dead black waters of a modern wastewetland. Later the draining or filling and polluting of wetlands by industrial technology increased markedly their degradation and destruction (for cities and wetlands, see Giblett 2016a).

Part of the problem with wetlands lies in the fact that they a taxonomic anomaly in a classificatory system, or "the order of things," predicated on a hard and fast distinction between not only land and water, but also time and space, or perhaps more precisely, their representational systems: the timelessness of maps and the spacelessness of history do not lend themselves to the changing nature of wetlands. What is needed, instead, are temporal maps that move with time, that show historical change in wetlands; spatial history that shows history taking place in, and in relation to, wetlands; temporal geography that shows natural processes occurring over time in wetlands; and quantum ecology that construes "the environment," especially wetlands, on a space/time continuum.

Part of the problem with wetlands lies also in the fact that they do not generally conform to the norms and conventions of aesthetics in the European landscape tradition. Aesthetics and wetlands have had a long and fraught relationship. Wetlands are generally not picturesque, nor beautiful, nor sublime (see Giblett 1996, 2020a). Yet these three are not the only aesthetic modes, or pleasures (or pleasure bordering on pain in the case of the sublime), possible. The fourth modality of the uncanny applies to the wetlandscape. The uncanny combines fascination and horror. It also engages the sense of smell with its immediacy, rather than the sense of sight with its mastery of the subject over the object (see Giblett 1992, 1993b, 1996, 2019b). The wetland is the uncanny place par excellence. The wetland is the home of the uncanny creatures par excellence of alligators and crocodiles (Giblett 2019b, chapter 2).

The wetlands are uncanny, combining fascination and horror. Both the uncanny and the wetlandscape involve developing an appreciation of smell quite inimical to the aesthetics of sight of the beautiful, picturesque, and sublime. Yet by way of dissociating the uncanny from its misogynist overtones (the ultimate uncanny place of horror for patriarchal culture is the mother's genitalia), it is necessary to make a distinction between the fascinatingly uncanny and the horrifically uncanny. Wetlands have by and large been the locus of the horrifically uncanny to shun, and destroy; they need to become a place of the fascinatingly uncanny to appreciate (its sights, sounds, and smells, even its tastes and textures) and preserve.

Wetlands have been, and still are to some extent, associated in mainstream western culture with disease, and with its possible, fatal outcome. They can even be associated not so much with the end of life as with lifelessness and the uninhabitable. This is not to deny that the swamp, or the wetland more generally, is a place of death—it is. But it is also a place of life. Wetlands are places of both life and death. They are living black waters. Death is necessary for life to be reborn. In the middle of death and decay in wetlands, we are in the middle of new life being reborn. Modern western culture has largely dismissed the association of wetlands with life, as living waters, and concentrated exclusively on the association of wetlands with despondency, despair and death, as black waters.

Despair has been associated with the color black, such as, for example, in Shakespeare's *Richard III* (as we will see in chapter 2 of the present volume). By contrast, Henry David Thoreau places hope in black waters (as we will see in chapter 8 of the present volume). Thoreau assists greatly with the construction of the central argument (of the present volume) against the mainstream western denigration of wetlands and makes the case compelling for their conservation. This case is -cultural—agricultural, aquacultural, horticultural, paludicultural—in a word, ecocultural, that is, "the cultures of nature" in the earthly household (see Giblett 2011, chapters 1, 2, and 12). The depths of hope in the slimy black waters of the swamp undermine the heights of despair in the sublime city and in the dead black waters of its slums and industrial wetwastelands. For Zoë Sofoulis (cited by Giblett 1996, 9 and 27), slime is the secret of the sublime that she encapsulates and conveys in her portmanteau word "s(ub)lime."

Besides being seen in mainstream western culture as places of disease, despair, depression, horror, and the monstrous, wetlands have also been, often as a direct result, as a refuge for the runaway and a site of resistance for the rebel or revolutionary against the dominant power. Wetlands are easy to defend and hard to attack. Many of the characteristics of the wetland that make it unattractive for the norm society, such as its darkness and impenetrability, are precisely those features which make it ideal for the rebel or revolutionary during time of war or suitable for the runaway from oppression during the time of "peace" of a sustained deprivation of liberty. The wetland is the locus of contradictory, mutually exclusive definitions, depending on whose point of view, even whose side, one takes, as is the definition of the rebel or revolutionary as either "terrorist" or "freedom-fighter" and patriot. Nowhere is this ambiguity more evident than when considering the role wetlands have played in military and slavery history where the same wetland can be, on the one hand, an intractable and inhospitable foe to the dominant but, on the other, a helpful and obliging friend to the dominated seeking refuge, and even a base for insurgency (see Giblett 1996, chapter 9).

EXPLORATION, COLONIZATION, AND URBANIZATION

When English explorers first came to Australia they brought with them the cultural baggage of the mainstream western denigration of wetlands. When they came to south-western Australia and the site for the future Swan River Colony (later the city of Perth) they were faced with an alien landscape (or more precisely wetlandscape) which did not conform to their preconceived ideas of what lakes should look like. So strange were the shallow, often seasonal wetlands of this new land that explorers did not mark them on their maps, nor even note the existence of lakes and swamps in their descriptions of the country (see Giblett 1993a, 1996, 2013a; Brady and Murray 2020; Ryan, Brady and Kueh 2020).

By contrast, when English explorers first came to eastern Australia and to Botany Bay, they noted the existence of mangrove swamps, but still thought it was a suitable site for the establishment of a settlement. When the first settlers came to eastern Australia they rejected Botany Bay in accordance with the miasmatic theory of disease in preference to Port Jackson, later the site for the city of Sydney. One of the reasons for choosing this site was the proximity of fresh water in the Tank Stream issuing from swampland in present day Hyde Park. When this source was polluted, settlers went further afield to a swamp in present day Centennial Park, which has been aestheticized and landscaped into a lake and lawns in accordance with the European landscape architectural aesthetic. When this source dried up, settlers went further afield to the swamps of Botany Bay, the very site they had rejected in the first place. Sydney had its birth in wetlands, but progressively killed them. The birth of Sydney and the death of its wetlands are taken up in chapter 6 of the present volume.

When English explorers first came to south-eastern Australia and Port Philip Bay, later the site for the city of Melbourne, they marked on their maps and noted in their descriptions of the country the existence of lakes and swamps. The Port Phillip wetlands were clustered around the shores of the Bay close to open, navigable water. They were readily accessible by English explorers circumnavigating within Port Philip Bay who made onshore excursions into the surrounding wetlands and went up the Yarra and Maribyrnong Rivers that debouched into deltaic marshes and swamps, whereas early Dutch, French and English explorers went up the nondeltaic Swan River and did not make onshore excursions into the inter-dunal Perth wetlands lying beyond.

When Captain James Stirling, an early explorer of the Swan River and later the first Governor of the Swan River Colony, wanted to sell real estate in the Colony he described the Swan Coastal Plain as "grassy country, thinly wooded," whereas in fact it was swampy country, thickly wooded. When it came to the foundation of a settlement, wetlands had to be mapped and were done so at a larger scale of a smaller area around the settlement of Perth, whereas the explorers' smaller scale maps of larger area wrote Perth's wetlands out of existence (Giblett 1993a, 1996, chapter 3; Brady and Murray 2020; Ryan, Brady and Kueh 2020). Wetlands to the north of the settlement were gradually drained or filled. Not until the 1990s were Perth's surviving wetlands in the suburbs and outskirts of the city precisely mapped in and as "A City of Wetlands." The birth of Perth and the life and death of its wetlands is taken up in these references.

The general view of settlers right into the twentieth century was that many Australian wetlands were not really lakes at all, and so not worthy of consideration, let alone conservation. They were also subject to the utilitarian view and found to be useless, though we now know that wetlands can even be more productive than rainforests. The project of colonization, especially in its modern phase and especially in relation to the establishment of settlements and the foundation of cities, is strongly tied to the draining or filling of wetlands; in fact, the latter makes possible the former. Without the draining or filling of wetlands the establishment and expansion of many modern cities would not have been possible. Some iconic cities of modernity, such as St Petersburg, were built on land reclaimed from marshes, while some new colonial settlements which grew into modern cities, such as Toronto (Giblett 2013b, 2014, 2016a), Perth (Giblett 1993a, 1996; Brady and Murray 2020; Ryan, Brady and Kueh 2020), and Melbourne (Giblett 2016b, 2020c) were founded between a river and wetlands. These and other wetland cities which became iconic cities of modernity, such as Paris, London, Venice, Berlin, Hamburg, Boston, New York, New Orleans, Washington, and Chicago, could only expand in an uninterrupted fashion by filling or draining the marshes or swamps.[10] The project of modernity with both industrial agriculture and the modern city is strongly tied to the draining or filling of wetlands. In fact, the latter made possible the former in many instances.

Melbourne is regularly voted "the world's most liveable city." It is also a UNESCO "city of literature," but it was, and still is, a city of wetlands, like Perth, though UNESCO does not have such a designation. Melbourne was the Australian capital of modernity (Giblett 2020c, chapter 1). In 2020 it has the dubious distinction of being the coronavirus capital of Australia. A trace of some of Melbourne's lost wetlands can be found in the areas peripheral to the Central Business District where they have been aestheticized and landscaped into lakes and parks. Some surviving wetlands can still be found in outlying suburbs. Urban dwellers of Melbourne are largely oblivious to the history of Melbourne's wetlands, but these wetlands are "present by virtue of their very oblivion" (to again use Benjamin's words), not least as "ghost swamps" (see Giblett 2020c, chapters 2–5). These swamps haunt the present consciousness of Melbourne and sit like a nightmare of history on the minds of the living (to use Marx's words) in "the now of recognizability" (to use Benjamin's words again) of "the dialectical image" of Melbourne as marsh metropolis. The birth of Melbourne and the death of its inner city wetlands are taken up elsewhere (Giblett 2016b, 2020c). A similar case applies to Paris (Giblett 2016, chapter 3).

Melbourne's lost wetlands are largely not commemorated and remembered in public installations or art works. Perth's lost wetlands are commemorated in memorials and were remembered in an exhibition mounted at Perth Town Hall in September 2014. The materials from this exhibition are now available online on the Western Australian Museum website (Brady and Murray 2020; Ryan, Brady and Kueh 2020). Until the recent exhibition by the Royal Historical Society of Victoria in early 2020 devoted to one lost wetland in Melbourne (as discussed in chapter 7 of the present volume), Melbourne's lost wetlands have not been commemorated in memorials, nor remembered in exhibitions, nor celebrated in public art works. To the best of my knowledge, a similar situation applies to Sydney's lost wetlands with no exhibitions, nor memorials, nor public art works devoted to them (as discussed in chapter 6 of the present volume).

When English explorers and settlers first came to North America (Canada and the United States) they brought with them the cultural baggage of the mainstream western denigration of wetlands (see Giblett 1996, 2014). They developed a farming frontier that involved not only clearing forests and grazing grasslands, but also draining and/or filling wetlands for the dryland agriculture of crops and grains. By contrast, when French explorers and settlers first came to Canada they not only developed a hunting and trading frontier, but also brought with them the agricultural technology for managing and farming wetlands (see Giblett 2014). North American colonial settlements established on or beside wetlands that developed into cities, such as Boston, New York, New Orleans, Toronto, Chicago, and Washington, followed the European model of the wetland city, such as Paris, London, Venice, Berlin, Hamburg, and St Petersburg, in draining, filling or canalising wetlands (see Giblett 2016a). North American wetlands and wetland cities are huge topics that have been discussed previously (Giblett 1996, 2014, 2016a). The American Henry David Thoreau's philosophy, psychology, and theology of swamps are discussed in chapter 8 of the present volume.

Wetlands have been colonized by the city, its settlers and its historians. The project of decolonization needs to include the decolonization of wetlands. Decolonization will not be achieved until wetlands are decolonized. Decolonization of wetlands has been addressed specifically in relation to Perth's wetlands (Giblett 1993a, 1996), to Melbourne's wetlands (Giblett 2016b, 2020c) and is in relation to Sydney's wetlands in chapter 6 of the present volume. Decolonization of Sydney's and Melbourne's wetlands could involve rehabilitating them, as well as acknowledging them by mourning their death and commemorating their life in memorials and other interpretation, such as in an exhibition or public art works.

The wetland has become literally ambivalent, a place of moving values, its meanings circulating and swirling around like its waters sometimes never settling on one definition, never flowing to an endpoint. At once the wetland is wasteland to be filled or drained, to be turned into profitable agricultural land or into a sanitary land-fill site and, at the same time, wilderness not yet subject to a capitalist imperative in which to seek refuge and from which to mount resistance to cultural and military imperialism. Against the rhetoric of swamps and marshes as places of melancholy and monstrosity, of horror and disease, a counter-tradition in the west has regarded the wetland as a sacred place, a place of both death and life. The counter-tradition sees wetlands as fully functioning ecosystems and habitats that harks back to the living black waters of pre-modern, pre-capitalist and pre-patriarchal (or matrifocal and gylanic) wetlands.

Present-day wetlands are the home—and often the only home—for wading birds (or "shorebirds," though wetlands often don't have shores, but are in dynamic transition from land to water). They are "birds of wetlands," "an ephemeral, marginal world" as Andrew Darby (2020, 2) puts it, so their home habitat is often marginalised and destroyed. They are transient, uncharismatic mid-scale fauna, yet they are also "astonishing ultramarathon birds" as Darby (2020) calls them, that fly annually back and forth "across the globe on a journey" seeking food, shelter, and breeding sites in the wetlands of the northern and southern hemispheres.

THE SWAMPY PATHWAY AFOOT

This section sets out the pathway through the rest of the present swampy volume by outlining the subsequent chapters. Chapter 1 of *Wetlands and Western Cultures* begins the body of the book proper by considering Christian theology as a foundation for the mainstream western cultural denigration of wetlands. This chapter considers wetlands in the Christian portrayal of wetlands in J. R. R. Tolkien's work in his scholarship on the Old English *Beowulf,* in *Beowulf* itself and in the second volume of his fantasy trilogy, *The Lord of the Rings*. These popular, canonical and scholarly texts engage uniquely with the theology of wetlands. They and John Milton's *Paradise Lost* are representative textual types of the Christian denigration of wetlands in literature and literary scholarship in mainstream western culture. These texts highlight the persistence of wetland denigration in similar terms as monstrous places of evil across a millennium from early medieval times in *Beowulf*, through early modern times in *Paradise Lost,* to late modern times in Tolkien's work.

In *Lord of the Rings* Tolkien's portrayal of the Dead Marshes drew from his experience of the trenches in World War I in which the muddy wasteland of the Western Front was the product of long-range artillery bombardment in modern industrial warfare. He applied the obvious denigratory features of the feral quaking zone of this culturally constructed and artificial wetwasteland to the native quaking zone of a natural marsh. Chapter 1 argues that Tolkien's denigration of wetlands in his scholarly work and fictional writing did not embrace *Genesis* 1: 1–2 in which God created the world as wetland, nor acknowledges *Genesis* 2: 6 in which water came up from the aquifers and watered the surface of the ground. Rather, in line with the Old English *Beowulf,* Tolkien emphasizes the wetland and its associated aquifers as a place created after the fall from grace out of which monsters are born and to which evil monsters are condemned.

Similarly, wetlands for Milton in *Paradise Lost* are the tomb in the underworld (rather than the womb of the world) where Satan as a swamp serpent and marsh monster was "engendered in slime" (as we will see in chapter 1 of the present volume). I posit that the Christian theology of wetlands, syncretized with the philosophy of the elements and humors by Milton (as we will see in chapter 1 of the present volume) and relayed by Bunyan (as we will see in chapter 2 of the present volume), is responsible for the destruction of wetlands over the past millennium. This conjunction between theology and philosophy formed a hegemonic alliance in the mainstream western denigration of wetlands for their destruction and against their conservation.

Deconstructing and decolonizing the Christian denigration of wetlands opens up space for wading into wetland cultures for the conservation of wetlands. Critiquing the environmental theology and politics of *Paradise Lost* and giving an ecocritical reading of it provides the opportunity to eco-creatively rewrite its denigration of wetlands in an environmentally friendly way. Chapter 1 constitutes the beginnings of a conservation counter-theology of wetlands taken up and elaborated in chapters 2 and 8 (and in Giblett 2021, chapters 18 to 29). The discussion of Christian literary perceptions of wetlands in these chapters does not include the history of other cultural perceptions, such as the agricultural, aquacultural, and paludicultural practices of livelihood using them from medieval to modern times, a very big topic and outside the scope of the present volume, much of which has been discussed elsewhere, including other literary perceptions (see Giblett 1996).

In contrast to the denigration of wetlands in the Christian theology of mainstream western culture typified by *Beowulf* and Tolkien's work and in Milton's syncretized philosophical theology, Freudian psychoanalysis provides a more productive path for the conservation of wetlands in developing psychoanalytic ecology, a talking cure for environ-mental illness and for nurturing environmental health. Chapter 2 of *Wetlands and Western Cultures* develops previous pioneering work in psychoanalytic ecology on wetlands. Psychoanalytic ecology, according to Esbjorn-Hargens and Zimmerman (2009, 523–524), "was first proposed in [. . .] *Postmodern Wetlands*" (Giblett 1996). They define it succinctly as "the application of Freudian and neo-Freudian concepts to the study of 'psychogeopathology' (i.e., the psychological underpinnings of destructive ecological behavior)." *Postmodern Wetlands* applied Freudian concepts of symptom, sublimation, mourning, melancholy, and the uncanny to the study of the psychogeopathology of the will to fill or drain and destroy wetlands. Neo-Freudian concepts of anal and oral sadism have been applied since and elsewhere to mining and pastoralism, and of psycho-symbiosis to living with the earth (Giblett 2011, chapters 9, 10 and 12; 2019d, chapters 5 and 6).

Chapter 2 of *Wetlands and Western Cultures* develops the previous discussion of mourning and melancholy in *Postmodern Wetlands* in relation to wetlands. It considers the psychology of wetlands by addressing the melancholy and mourning often associated and experienced with them. Wetlands were once thought to be bad for the body under the miasmatic theory of disease that malaria, literally "bad air," could be caught be breathing miasma or bad air. Wetlands could also be bad for the mind. Indeed, they can plunge the mind into melancholia, and even into madness, into despondency and even into despair. Chapter 2 begins by considering John Bunyan's "slough of despond" in *The Pilgrims Progress.* Despondency is another name for melancholy that is the immanent counterpart to the transcendental sublime and the spiritual counterpart to the psychological uncanny. As the sublime is a secular theology for a world in which God is dead and as the uncanny is a secular demonology for a world in which Satan is dead, so melancholy is a secular spirituality in the lower psychopathological register of what Julia Kristeva (1982) calls "the abject," the mediating category between subject and object that makes both possible. Melancholy is associated with death, or at least with the valley of the shadow of death (the precursor to death) and with mourning (the aftermath of death). Melancholy and mourning are connected for Freud in terms of loss: loss of the self or loved one. Chapter 2 goes on to consider Freud's essay on the topic, E. M. Cioran's discussion of melancholy and despair, Charles Dickens' depiction of "the grim domains of Giant Despair," Ilit Ferber's reading of Walter Benjamin's work on melancholy in the German mourning play, and Vera Brittain's melancholic and mournful account of her visit to the swampy city of Venice in the aftermath of the feral quaking zone of the muddy Western Front of World War I.

Chapters 1 and 2 of *Wetlands and Western Cultures* comprising part 1 are thus devoted to discussing wetlands and their -ologies: the theology, counter-theology, psychology, psychogeopathology and psychoanalytic ecology of wetlands. Taken together, they constitute a critical conservation wetland-ology, or aquaterratology, to counter and resist the misaquaterrism, the hatred and denigration of wetlands, of mainstream theology and psychology, and aquaterracide, the killing of wetlands, or genocide of wetlands by mainstream western culture. Aquaterratology is ecology "with nature" (see Mules 2014). It enables and nurtures "living with the earth" in what I call "bio- and psycho-symbiotic livelihoods in bioregional home habitats," including wetlands (see Giblett 2011). Aquaterratology is a new trans-discipline studying wetlands that not only critiques and bridges Greco-Roman and Judeo-Christian thought and culture (as the name implies and encapsulates), but also crosses the great divide between the "two cultures" of the sciences and the humanities (and gets shot down in the "no man's land" or wasteland of the feral quaking zone between them that few dare to cross).

The present Introduction outlines the theoretical framework and political platform, and chapters 1 and 2 provide the counter-theological and psychoanalytic ecological basis, for the critical discussion of the case studies of three geographical areas and three cities in chapter 3–7. This basis, framework and platform are also developed and elaborated in and by these chapters, rather than simply and mechanically applied to, or merely illustrated by, the discussion of these areas and cities. *Wetlands and Western Cultures* also builds on the foundation of the work of many others, especially Zoë Sofoulis, as acknowledged, referenced and used elsewhere.[11] The penultimate chapter 8 shows how Henry David Thoreau, "the patron saint of swamps" (Giblett 1996, ix, 229; 2020d, chapter 4), began this psychological, ecological and theological thinking about and with wetlands. This thinking is aquaterratological, or what Ryan and Chen (2020) call "swamp thinking," thinking about and with swamps. The concluding chapter 9 shows how I tried to follow in Thoreau's big swampy footsteps with my own nature writing. The "Preface" to the present volume has shown the development of my own thinking and writing about and with wetlands.

Wetlands have been denigrated in and by mainstream western culture partly because they do not conform to the conventions of European landscape aesthetics and its three main modes of the sublime, the beautiful ,and the picturesque (Giblett 1996, chapter 2). Wetlands can look ugly and smell bad so they have been devalued. Whatever looks bad or smells bad must be bad morally. Part 2 of *Wetlands and Western Cultures* comprising chapters 3–5 considers wetlands in art, history, aesthetics, landscape painting, environmental artwork, and literature by presenting three case studies of wetlands and their arts and cultures in three geographical areas of England and Australia. These three chapters show the persistence of the mainstream western cultural view and the resistance to it by tracing the shift from denigration to conservation in three widely separated locales. They build on previous work on North American and Australian wetlands and wetland cultures (Giblett 1996, 2014, 2016, 2020a).

Chapter 3 of *Wetlands and Western Cultures* considers the English Fens in politics, painting, poetry, prose, and art history. Recent work on the English Fens coming out of art history, landscape aesthetics, and landscape architecture in Vittoria di Palma's (2014) *Wasteland* demonstrates the still contentious and problematic positioning of the Fens in English culture and the need for concerted efforts to conserve them. This chapter of *Wetlands and Western Cultures* takes a political approach to the Fens and to the enclosure of the Commons (including much of the Fens) into private property critiqued by John Clare, Karl Marx, E. P. Thompson and Raymond Williams. It also takes a cultural, literary, and historical approach by considering the depiction of the Fens in painting and the literature of the Fens, beginning with Clare's poetry and prose in the nineteenth century for the Fens and against enclosure. It then proceeds to a consideration of the Fens in later prose, beginning with Dorothy Sayers' classic detective novel *The Nine Tailors* (1934), proceeding to Graham Swift's wetlandmark *Waterland* (first published in 1983) and his 25th anniversary introduction of 2008, Robert Macfarlane's (2008) "new nature writing" of "Ghost Species," and culminating recently with Becky Owen-Fisher's poetry (2017) and Stella Tillyard's historical romance *The Great Level* (2018).

The discussion in this chapter of the artistic and literary perceptions of the Fens does not include a history of the Fens and other cultural perceptions of them, such as the agricultural, aquacultural, and paludicultural practices of livelihood using them, a very big topic and outside the scope of the present volume, some of which has been covered elsewhere, beginning with the pioneering work of H. C. Darby dating from the 1950s (as referenced and discussed in Giblett 1996). The discussion of the Fens in this chapter is also not an ethnographic study of local stories. To gather such stories would require resources and opportunities outside my ability to access.

Chapters 4 and 5 of *Wetlands and Western Cultures* go on to discuss wetlands in some instances of Australian landscape aesthetics and painting and environmental artwork beginning in chapter 4 with a case study of the wetlands of the Gariwerd area in western Victoria. In the 1830s the explorer/surveyor Major Mitchell dubbed this area "Australia Felix," meaning "Australia the blessed" because of its pleasing prospects (in two senses, spatial and temporal) for pastoralism. Not surprisingly, this area was also a happy hunting ground for heroic settler-colonial landscape painters, such as Eugene von Guérard. This chapter follows in the swampy footsteps of "Artist and Swamp: Wetlands in Australian Painting and Photography" (Giblett 2020a). Gariwerd, or "the Grampians" as Mitchell renamed them, is largely known collectively as the mountainous region in western Victoria centered around the town of Halls Gap nestled in the valley between two north-south ranges and accessible from the east through a gap (hence the name) in the ranges. Hardly surprisingly also, the water flowing off these spectacular ranges also created magnificent wetlands around the base of the ranges. The Grampian mountain ranges running north-south are a watershed with river catchments to the east and west feeding rivers flowing to the north and south, and creating wetlands. The Grampian mountain ranges are "the head," or top, of these catchments.

Gariwerd, or the Grampians, are not known for, or associated with, the wetlands that encircle the mountain ranges like a necklace of sapphires as depicted in one of von Guérard's paintings. These wetlands nevertheless played an important role in the lives of Aboriginal people and English and Scottish settlers. Chapter 4 of *Wetlands and Western Cultures* weaves together the strands of local and environmental history with Aboriginal story. Illustrated with landscape paintings by von Guérard and with environmental artwork by local artist Carole Mules in mixed fabric collages creating textured renderings of the wetlands of western Victoria, it celebrates the life of the wetlands around the mountain ranges of the Grampians in western Victoria and their role in the lives of the people who have called this place home for many generations. Based on archival and contemporary research into Grampians wetlands, this chapter conserves and conveys the natural and cultural heritage of the area for present and future generations. By the by, it also solves the mystery of the messiah of conservation who leads the twelve environmental apostles as assembled and introduced in *New Lives of the Saints* (Giblett 2020d).

Chapter 5 of *Wetlands and Western Cultures* continues the discussion of wetlands in landscape aesthetics and environmental artwork in Australia turning to a case study of the Stirling Range in Western Australia and its wetlands. By contrast with Major Mitchell's pleasing pastoral prospect of "Australia Felix," the Stirling Range could be called "Western Australia Felix," Western Australia the blessed, because of its stunning botanical biodiversity with more species of native flora than in the entire British Isles. The area was devastated by a bushfire in late 2019 caused by lightning strikes and exacerbated by global heating. One of the major plant communities contributing to and supporting its biodiversity is its wetlands. However, despite the government botanist documenting the huge plant diversity of the Stirling Range from the 1840s, a former professor of geology at the University of Western Australia effectively described the surrounding area in 1920 as virtually "Western Australia Execratus," Western Australia the cursed, because of its sandy plains, salt lakes, and stunted vegetation. The only redeeming feature of the Range for him was one singular, sublime mountain. He and a recent coffee book were singing from the song-sheet of the European landscape aesthetic of the sublime that valorised mountains and denigrated wetlands. As with the previous chapter, chapter 5 of *Wetlands and Western Cultures* weaves together the strands of local and environmental history with Aboriginal story and, illustrated with the recent environmental artwork of Joanne Francis, it celebrates the local wetlands as wombs of new life and tombs of old life in which new life is reborn, especially challenging after the recent bushfire. Based on archival and contemporary research into the Stirling Range wetlands, this chapter conserves and conveys the natural and cultural heritage of the area for present and future generations.

Wetlands are wombs of new life and tombs of old life out of which new life springs (Giblett 1996, 163, 176). In a word, wetlands are places and spaces of what Hannah Arendt (1996, 51, 132–133, 146–148), the eminent philosopher and the student and lover of Martin Heidegger, called "natality," a term she coined in her doctoral thesis on love in the late 1920s.[12] Arendt placed natality, or being born, at the centre of her social philosophy of life and love as a counter to Heidegger's morbid, solipsistic philosophy of mortality and despair facing the fact of one's own death. Her philosophy of being born (and born again) counters his philosophy of being towards death. Death is a part of life, but new life is born in an eternal return and cycle of birth, life, death, birth, etc.

Arendt also placed natality and being born again into the public sphere (*polis*) out of the private sphere of the domestic household (*oikos*) at the centre of her political philosophy. Wetland conservationists, such as Henry David Thoreau (as we will see in chapter 8), place being born again in swamps at the centre of their philosophy, psychology and theology of aquaterratology and at the heart of their practice of wetland conservation in aquaterranatality. From aquaterratology to aquaterranatality marks the conceptual and political pathway, compositional structure, and presentational trajectory through the slimy depths of the present swampy volume.

Rather than the -ologies (theology, counter-theology, psychology, psychogeopathology, psychoanalytic ecology, aquaterratology, etc.) that privilege the logos or word and the -isms in general that exploit and destroy earth (industrialism, capitalism, neo-liberalism, etc.), and sanctuarism in particular that only cares for special places while the rest of earth goes to hell in a handbasket, sacrality and aquaterranatality in particular and the -ities in general (mutuality, materiality, spirituality, generosity, etc.) care for earth, wetlands and all earthly places, including cities (for the "-isms" and "-ities," see Giblett 2011, especially figure 2, 32–34).

Wetlands and cities, wetlands in cities, have had a longstanding fraught relationship with many cities draining or filling wetlands so that the city could be built in the first place. Many cities grew out of colonial settlements established in or by wetlands (Giblett 2016a, b). Chapters 6 and 7 comprising part 3 of *Wetlands and Western Cultures* present case studies of wetlands and water in three cities in England and Australia, all sites of colonization of wetlands and icons of modernity: Sydney, London, and Melbourne, including a discussion of, and proposals for, wetlandscape architecture in the case of the latter two cities, whereas Sydney has neglected conserving and commemorating this aspect of its culture, geography and history. A recent history of water in Australia acknowledges that for the first colonial settlement of Sydney the nearby wetlands were an important source of water, but they were quickly degraded, destroyed and largely forgotten. The same situation applies with Perth and Melbourne with wetlands initially providing water for humans and stock, feed for the latter, and then they were degraded and destroyed. These two chapters build on previous work on North American and Australian wetland cities (Giblett 1996, 2014, 2016, 2020c).

Recent wetland cultural studies in the environmental humanities on Australian cities, such as the birth of Perth (Brady and Murray 2020; Ryan, Brady and Kueh 2020) and Melbourne (Giblett 2020c), acknowledge the wetlands that were present at settlement and are now absent as cities of ghost swamps, mourn the loss of these wetlands, call for the conservation of those that remains, propose memorials for those that have been lost, and interpretation of the history and significance of all of them. Chapter 6 proposes that the same courtesy be extended to the largely forgotten wetlands of Sydney. The birth of these three Australian cities near wetlands meant the death of those early wetlands that were close to colonial settlements. Remnants of wetlands can be found further afield on the outskirts of all three cities and in inland areas of all three states of which they are the capital city.

In recent work on Melbourne and London, water is seen, carried, transported, disposed of, designed for, and so on, in and for these two cities in four different paradigms in the disciplines, discourses and practices of: urban planning (Dovey and Jones 2018; Jones 2018); landscape architecture (Bertram and Murphy 2019); urban geography (Gandy 2014); and wetland cultural studies in the transdisciplinary environmental humanities (Giblett 1996, 2008, 2016a, 2020c). Chapter 7 considers each of these aspects in turn. It also considers wetlandscape architecture that constructs artificial wetlands where land and water are in transition from one to the other. Wetlandscapes should be regarded as legitimate activities and places of landscape design and should not regarded as second-class citizens to first-class parklandscapes. The discussion in these chapters of water and wetlands does not include a history of these cities and their wetlands, a very big topic and outside the scope of the present volume, much of which has been covered elsewhere.[13]

In sum, wetlands have been denigrated in and by mainstream western (as discussed in the present volume unless indicated otherwise):

1. theology (of creation and demonization of the monstrous);
2. philosophy (of the elements as an aberrant mixture of earth and water);
3. psychology (of the humors as melancholic);
4. medicine (of miasmatic malaria; see Giblett 1996);
5. aesthetics (of the sublime, the beautiful and the picturesque; see Giblett 1996);
6. landscape painting and photography (of the aesthetic; see Giblett 2011, 2020a);
7. exploration (of foreign lands); and
8. literature (with the above, including explorers' journals; see also Giblett 1996).

Denigration led to, or went hand-in-hand with, destruction; destruction was underpinned and justified by denigration.

Wetlands have been destroyed in and by mainstream modern western:

1. engineering (by industrial-scale draining, dredging, and filling);
2. agriculture (by industrial-scale dryland farming);
3. pastoralism (by industrial-scale sheep and cattle ranging);
4. cadastral mapping and surveying (for settlements, e.g., grid plan town);
5. town planning and urban development (by city and suburb construction);
6. architecture and building construction (with brick, stone, steel, and concrete);
7. landscape architecture, and parks and garden construction;
8. warfare (by bombardment [aerial and artillery]); and
9. colonization (by all of the above).

Resisting and countering mainstream western culture's denigration and destruction of wetlands, they have been celebrated and conserved in and by (as discussed in the present volume unless indicated otherwise):

1. indigenous cultures (e.g., Australian Aborigines; see Webb 1996 and Giblett 2020e);
2. political resistance (e.g., Fenlanders);
3. counter-theology (e.g., Henry David Thoreau on swamps);
4. counter-aesthetics (of the uncanny, see Giblett 1996, chapter 2; 2019d; also for example Thoreau);
5. evolutionary biology (e.g., Thoreau);
6. environmental humanities (e.g., wetland cultural studies, psychoanalytic ecology; aquaterratology; see also Giblett 1996, 2019d);
7. wetlandscape architecture and design (e.g., Nigel Bertram and Catherine Murphy);

8. wetlandscape photography (e.g., Simon Neville in Giblett and Webb, eds 1996);
9. environmental artwork (e.g., Carole Mules and Joanne Francis);
10. nature writing (e.g., John Clare on the Fens, Thoreau); and
11. wetland conservation (aquaterranatality; Mules, Francis and Thoreau).

Wetlands and Western Cultures concludes in part 4 with a consideration of nature writing advocating (rather than merely arguing for) the conservation of wetlands, beginning in chapter 8 with a discussion of the pioneering work of Henry David Thoreau from the mid-nineteenth century. Thoreau is famous as a nature writer and environmental philosopher of the wilds and the woods, as an early advocate of evolutionary biology and as a pioneer of forest ecology. He reflected and wrote philosophically on bogs, marshes, and swamps. He has thus been dubbed the "patron saint of swamps" (Giblett 1996, ix, 229; 2020d, chapter 4), though most writers about Thoreau, including all his biographers, neglect or overlook this aspect of his writings. Chapter 8 of *Wetlands and Western Cultures* aims to remedy the neglect or oversight of this aspect of his work. It also partly aims, not only to supply this lack, to fill this absence, but to show the importance of Thoreau's writings for critiquing the conventional cultural construction and perception of wetlands in mainstream western culture. He upset and overturned mainstream thinking about swamps, including the conventions of European landscape aesthetics, philosophy, medicine, theology, and psychology that regarded them as dismal, disease-ridden, and displeasing. He also developed a conservation counter-aesthetic and a counter-Christian religion of swamps. He valued swamps as quaking zones where body and mind meet in the premodern wetland. As a proponent of evolutionary biology and as a non-believer in the institutionalized religion of Christianity, he developed his conservation counter-theology against creationism and for evolution in swamps.

In conclusion, chapter 9 of *Wetlands and Western Cultures* discusses *Black Swan Lake* (Giblett 2013a) in the context of the nature writing of Henry David Thoreau on swamps and Aldo Leopold on marshes. Tracing the life of the plants and animals of Forrestdale Lake in Western Australia through the six seasons of the local indigenous people, the first part of *Black Swan Lake* presents a wetlands calendar over a yearly cycle of the rising, falling, and drying waters of this internationally important wetland. The second part of the *Black Swan Lake* considers issues and explores themes from the first part, including a cultural history of the seasons and the black swan. *Black Swan Lake* is a book of nature writing and environmental history and philosophy arising from living with other beings in a particular place. It is a guide to living simply and symbiotically with the earth in troubled times and places by making and maintaining a strong attachment and vital connection to a local place and its flora and fauna. Local places and their living processes sustain human and other life on this living earth. I built my own house by Forrestdale Lake in 1986 and lived in it for 28 years. I kept a nature journal for a couple of years from which the first part of *Black Swan Lake* is drawn. Chapter 9 of *Wetlands and Western Cultures* concludes with a plea for living bio- and psycho-symbiotic livelihoods in bioregional home habitats of the living earth with their living black waters (Giblett 2011, chapter 12).

Wetlands and Western Cultures is a cultural, political, and environmental intervention for the conservation of wetlands around the world. It not only critiques the conventional positioning of wetlands in mainstream western culture, but also provides positive and affirmative ways of thinking, living, and being with wetlands. It invites the reader to wade into wetlands like a wader bird, to continue on its swampy pathway from aquaterratology to aquaterranatality and to participate in the conservation of wetlands. Black waters live!

NOTES

1. For the cultural and environmental history of the color black, see Giblett (2013a, chapter 14).

2. Matrifocal is focused on the maternal; gylany is the social structure in which all the sexes are equal; and including see Giblett (1996, xi, xii n.1).

3. "Mainstream western culture" used in the present volume is shorthand for, and an umbrella term covering, predominantly Anglophone cultures, Greco-Roman philosophy and Judeo-Christian theology. The scope of the present volume is regrettably limited to these aspects and to the tradition counter to them within western cultures. Other aspects of mainstream western culture, philosophy and theology, and other geographical areas and histories, invite possible application of this approach and further elaboration, comparison and contrast. Some cross-cultural comparison with Australian Aboriginal cultures of wetlands is made periodically; see also n.7 below. The present volume is thus not a comprehensive discussion of wetlands and western cultures *per se*, nor of wetland cultures in all western countries, both of which are huge topics, would require multiple volumes and are outside my ability and expertise to research and write. Some American wetlands, such as the Great Dismal Swamp, are discussed in Giblett (1996); some Canadian wetlands are discussed in Giblett (2014); and some other Australian wetlands are discussed in Giblett and Webb (1996).

4. For further discussion of slime, see Giblett (1996, chapter 2).

5. For further discussion of the sublime, see Giblett (1996, chapter 2).

6. For a pictorial portrayal of some of these transformations in what I call "A Psychogeocorpography of Modernity," see figure 3.1 in the present volume. For the sublime city drawing on the work of Norman O. Brown, see Giblett (2016a, 131).

7. I discuss wetland commons in chapter 3 of the present volume.

8. This topic is outside the scope of the present volume and is discussed previously by myself and others. For further discussion see, Giblett and Webb (1993, 1996); Webb (1996); Giblett (1996, 2020e).

9. For the concept of the Symbiocene, see Glenn Albrecht (2019; see also Giblett 2016a, 12 and 251, n.1). I differentiate the two paradigms of the drive for mastery and the desire for mutuality from each other in Giblett (2011, chapter 1). I make the argument and call for living bio- and psycho-symbiotic livelihoods in bioregions of the living earth in Giblett (2011, especially chapter 12). I discuss and differentiate the Anthropocene and the Symbiocene in the final chapter of the present volume.

10. All these cities, as well as St Petersburg and Toronto, receive a chapter-by-chapter treatment in *Cities and Wetlands* (see Giblett 2016a).

11. Beginning with Giblett (1996) and in the "Acknowledgments" of the present volume. For a pictorial depiction of this theoretical framework, see figure 3.1 in the present volume.

12. For Arendt's philosophy of love and her relationship with, and views of, Heidegger, see Kristeva (2001, xv, xvii, 8, 44–46, 48, 66–67, 141, 239–240, especially 243 n27 for Arendt's outrageous and scathing "true story of Heidegger the fox," 246 n75) and Eilenberger (2020, 192, 240, 379 n22, 24, plus photos of Heidegger, the sly fox).

13. For London, see Giblett (2016a, chapter 5); for Melbourne, see Giblett (2016b, 2020c, chapter 2–5). To the best of my knowledge, an extensive history of Sydney and its wetlands remains to be researched and written.

Part I

WETLANDS AND -OLOGIES

Chapter 1

Theology of Wetlands

Tolkien, Beowulf, *and Milton on Marshes and Their Monsters*

From the Old English *Beowulf* to J. R. R. Tolkien's work, whether as the writer of *Lord of the Rings*, or as critic and scholar of Old and Middle English literature, wetlands and marsh monsters get bad press. The Old English *Beowulf* interpolates several verses into the biblical book of *Genesis* that explain the propagation of monsters and that justify regarding wetlands as evil, pagan places. Tolkien's view of marshes and marsh monsters developed partly out of the Old and Middle English and Christian literary precursors he is drawing on in *Beowulf* and *Sir Gawain and the Green Knight*; partly out of his Christian beliefs about wetlands theologized and moralized as places of evil and monsters, a view forged in concert with his critical and scholarly work on *Beowulf* and *Sir Gawain and the Green Knight* and expressed in his portrayal of the Dead Marshes in *Lord of the Rings;* and partly out of his personal experience of the wet wasteland of mud and slime on the Western Front in World War I, the prototype for the Dead Marshes in *The Lord of the Rings*.

These popular, canonical, and scholarly texts engage uniquely with the theology of wetlands. They and John Milton's *Paradise Lost* are representative types of the Christian denigration of wetlands in literature and literary scholarship in mainstream western culture. These texts highlight the persistence of wetland denigration in similar terms as monstrous places of evil across a millennium from early medieval times in *Beowulf*, through early modern times in *Paradise Lost*, to late modern times in Tolkien's work.

In *Lord of the Rings* Tolkien's portrayal of the Dead Marshes drew from his experience of the trenches in World War I in which the muddy wasteland of the Western Front was the product of long-range artillery bombardment in modern industrial warfare. He applied the obvious denigratory features of the feral quaking zone of this culturally constructed wetwasteland to the native quaking zone of a natural marsh. In this chapter I read and critique the portrayal of wetlands in Tolkien's work and argue that his Christian reading of them did not embrace *Genesis* 1: 1–2 in which God created the world as wetland, nor *Genesis* 2:6 in which water came up from the aquifers and watered the surface of the ground. Rather, in line with *Beowulf,* he emphasizes the wetland and its associated aquifers as a place created after the fall from grace out of which monsters are born and to which evil monsters are condemned.

Wetlands for Milton in *Paradise Lost* are the tomb in the underworld (rather than the womb of the world) where Satan as a swamp serpent and marsh monster was "engendered in slime." I suggest that the Christian denigration of wetlands syncretized with the philosophy of the elements and humors by Milton (as we will see later in the present chapter) and relayed by Bunyan (as we will see in chapter 2 of the present volume), is responsible for the destruction of wetlands for the past millennium. This conjunction between theology and philosophy formed a hegemonic alliance in the mainstream western denigration of wetlands, for their destruction and against their conservation. Deconstructing and decolonizing this denigration opens up space for wading into wetland cultures for the conservation of wetlands. Critiquing the environmental politics of *Paradise Lost* and giving an ecocritical reading of it provides the opportunity to eco-creatively rewrite its denigration of wetlands in environmentally friendly way. Chapter 1 constitutes the beginnings of a conservation counter-theology of wetlands taken up and elaborated in chapters 2 and 8 (and in Giblett 2021, chapters 18 to 29).

Tolkien's portrayal of wetlands in these works presents a mode of thinking about the wetland world that is conservatively in line with dominant cultural discourses that is Christian in its genesis, but out of line with modern scientific, pre-modern sacral, and postmodern countercultural views of wetlands—perhaps unlikely bedfellows. Wetlands are vital for life on earth, including human and nonhuman life. The leading intergovernmental agency on wetlands states that:

> They are among the world's most productive environments; cradles of biological diversity that provide the water and productivity upon which countless species of plants and animals depend for survival. Wetlands are indispensable for the countless benefits or "ecosystem services" that they provide humanity, ranging from freshwater supply, food and building materials, and biodiversity, to flood control, groundwater recharge, and climate change mitigation. Yet study after study demonstrates that wetland area[s] and [their] quality continue to decline in most regions of the world. As a result, the ecosystem services that wetlands provide to people are compromised. (Ramsar Convention Bureau, n.d.)

Yet more than the mere providers of "ecosystem services," like sexual services, wetlands are habitats for plants and animals, and homes for people. Of course, it is churlish to critique the mistakes of the past from the privileged vantage point of the present. Yet rather than merely bemoan the acts of the past and the facts of history, the point is that some of the mistakes of the past are perpetuated into the present with the continued destruction of wetlands supported by the denigration of wetlands in the works mentioned.

By contrast with Tolkien's and Milton's denigration of wetlands, I rewrite passages from *Paradise Lost* from an environmentally friendly point of view. I also conclude this chapter by presenting the recent view of Paul Kingsnorth in *The Wake* in which the wetland is a place of burgeoning and decomposing life, of both life-giving and death-dealing water, of sacrality and resistance to hegemony. *The Wake* was published in 2014, the same year in a strange coincidence as the posthumous publication of Tolkien's translation of *Beowulf*. Diana Gabaldon's *An Echo in the Bone* is also another recent instance of a similar view of wetlands with her account of "The Great Dismal" Swamp (Gabaldon 2009, 369–374).

BEOWULF'S MONSTROUS FENS

Tolkien's Christian denigration of wetlands was forged in concert with his reading and translation of, and in his remarks and commentary on, *Beowulf*, described by Tolkien (2006, 30–31) himself as "the most successful Old English poem," more specifically as "an heroic-elegiac poem," and so not an epic poem, though it includes the obligatory descent of the hero in epic poems into the underworld, often a wetland as in this poem. In *Beowulf* Hrothgar describes Grendel the monster and his mother as two:

> huge wanderers of the marches guarding the moors, alien spirits [. . .] They dwell in a land unknown, wolf-haunted slopes, wind-swept headlands, perilous marsh-paths, where the mountain stream goes down under the mists of the cliff,—a flood under the earth. It is not far hence, in miles, that the lake stands over which hang groves covered with frost: the wood, firm-rooted, overshadows the water. There may be seen each night a fearful wonder,—fire on the flood! [. . .] That is no pleasant spot. Thence rises up the surging waters darkly to the clouds, when the wind stirs up baleful storms, until the air grows misty, the heavens weep. (*Beowulf* 1950, 89–90)

Seamus Heaney (*Beowulf* 1999, 45) in his translation prefers "uncanny" to "a fearful wonder," quite aptly in my view as the wetland is the uncanny place par excellence (as I have argued elsewhere; see Giblett 1996, chapter 2). It is "not far hence in miles" because it is closer than you think in mind. This is a typical trope for the horror of swamps.

Hrothgar's description of this wetlandscape can be compared with a similar, later Medieval one in *Sir Gawain and the Green Knight* translated by Tolkien as "the heavens are lifted high, but under them evilly/mist hangs on the moor, melts on the mountain" (*Sir Gawain* 1975, 77). The wetland in both texts not only mixes the elements of water and earth, and air and water when the atmosphere is misty in the temperate zone (and dripping wet in the tropics), but also mixes fire and water when marsh gases are ignited in the temperate zone. The traditional dragon breathing fire is arguably just a marsh (monster) "breathing" ignited marsh gases. Fire is also associated with hell, with the burning fires of brimstone, so the wetland in patriarchal western culture is a kind of watery hell.

Both Grendel and his mother are alien spirits because they are "Cain's kindred" (*Beowulf* 1950, 25) condemned by God to the wetland hell. The worst of the two monsters, mother and son, is the son, "that grim spirit [. . .] called Grendel, the renowned traverser of the marches, who held the moors, the fen and fastness; unblessed creature, he dwelt for a while in the lair of monsters, after the Creator had condemned them." Grendel, however, has not only been condemned but also been made into the personification, or more precisely "monstrification," of damnation. He is described as "the grim and greedy creature of damnation, fierce and furious," and as "the demon, the dark death-shadow" who "in the endless night [. . .] held the misty moors" (*Beowulf* 1950, 26 and 28). Grendel is a demonic agent, or "angel," of death and an orally sadistic monster who lives in the wetland realm of darkness.

Grendel's mother is not much better as she is described as "monstrous among womenkind [. . .] who must needs inhabit the dread waters, chilling stream" (*Beowulf* 1950, 84). Grendel's mother lives in the dreadful aquifer, the flood beneath the earth, which is even worse than the flood on or over the earth. Grendel's mother is an instance of what Barbara Creed (1986, 44–70) calls "the monstrous-feminine," or more precisely the monstrous maternal, a patriarchal figuring of the mother as monster that denies and represses the gylanic Great Goddess or Mother of the sacral swamps and marshes (see Giblett 1996, 2011). In patriarchal Western culture the pejoratively coded monstrous maternal marsh and maternal marsh monster by no means began with *Beowulf*, nor ended there (see Giblett 2008a, chapter 5). They go back to classical Greek times with, for instance, the Herculean labor of killing the monstrous Stymphalian birds that live in a marsh. Grendel and his mother are both specifically marsh monsters and a part of a long genealogy in patriarchal western culture.

Both are also orally sadistic marsh monsters with Grendel variously described as "the grim and greedy creature of damnation" (*Beowulf* 1950, 26), as "greedily loping" (*Beowulf* 1999, 24) and as "the monster" who "seized a sleeping warrior" and "rent him greedily" (*Beowulf* 1950, 58).

Grendel is moreover "a devouring murderer" who "consumed the beloved man's whole body" (126). Heaney's translates this line in part as "Grendel's maw" (*Beowulf 1*999, 66). "Maw" is a pejorative term for mouth often used in relation to orally sadistic monsters, such as alligators, crocodiles, and sharks. Grendel's mother is also described as "fiercely ravenous" (*Beowulf* 1950, 96), translated by Heaney as "gluttonous" (*Beowulf* 1999, 49) and by Clark Hall as "greedy and grim" (*Beowulf* 1950, 96). Orally sadistic monsters, I have argued elsewhere, are projections, displacements, and disavowing devices for the greed and gluttony meted out to the earth in mining, pastoralism, and wetlands dredging and draining, especially as carried out later by orally sadistic monstrous machines that consume the earth (see Giblett 1996, 2011).

In *Beowulf* Cain, the biblical son of Adam and Eve, and brother and murderer of Abel, is regarded as the progenitor from whom "all evil broods were born, ogres and elves [Tolkien (2014, 16 and 162) has 'goblins' as elves are good in his view] and evil spirits—the giants also, who long time fought with God" (*Beowulf* 1950, 25). In his lecture on *Beowulf* Tolkien (2006, 26) states that Cain is "the ancestor of the giants." Tolkien (2014, 161–163) in his recently published commentary on *Beowulf* claimed that this specifically Christian passage (but one that has no biblical warrant or foundation) is an addition or insertion by the Beowulf poet into the older material he is assimilating for this specific composition. Later in *Beowulf* (1950, 85) the poet says that from Cain "were born numbers of fateful spirits, of whom Grendel was one, a hateful outcast foe [. . .] the monster." Grendel was descended from Cain by way of Grendel's mother. The *Beowulf* poet effectively interpolates a concocted confection in *Genesis* chapter 4 between verses 16 and 17 and after the story of Cain's murder of Abel:

> 16 Then Cain went away from the presence of the Lord, and settled in the land of Nod, east of Eden. (English Standard Version)
>
> 16a Thence all evil broods were born, ogres and goblins and evil spirits—the giants also, who long time fought with God.
>
> 16b From Cain were born numbers of fateful spirits, of whom Grendel was one, a hateful outcast foe, the monster.
>
> 16c That grim spirit called Grendel, the renowned traverser of the marches, held the moors, the fen and fastness; unblessed creature, he dwelt for a while in the lair of monsters, after the Creator had condemned them.
>
> 17 Cain knew his wife, and she conceived and bore Enoch; and he built a city, and named it Enoch after his son Enoch. (English Standard Version)

Cain gave birth to monsters in the land east of Eden, the land outside the Garden of Eden. Monsters were thus not born in Eden, the sinless prelapsarian garden of God, but outside it in the wetlands before the city both temporally and spatially. Cain also gave birth to the city (the postlapsarian city of sin, the monstrous city) after he gave birth to monsters. The city came after God had completed His creation, including the Garden of Eden, after the Fall and then, after Cain murdered Abel, God cursed Cain and Cain created the city.

Cain is figured as giving birth like Zeus to monsters in perhaps the first recorded instance in English literature of a bachelor birth, though unlike Zeus who gave birth to Pallas Athene out of his head, there is no mention as to the part of his body from which Cain gave birth to monsters. Just as the birth of Pallas Athene from the head of Zeus after he had ingested her mother Metis is a Greek patriarchal myth of origins that denies or represses and appropriates the role of the human mother in giving birth to children so Cain's giving birth to monsters is a Christian patriarchal myth of origins that denies or represses and appropriates the role of the great mother of the marshes in giving birth to wetlands and its creatures. Although there is certainly no biblical warrant or basis for regarding monsters as Cain's kindred and him as their progenitor, Tolkien (2006, 19) argued in his lecture on *Beowulf* that these references are vital for the poem, or in his terms, that in them "the key to the fusion-point of imagination that produced this poem lies."

The creator certainly condemned Cain for murdering his brother (*Genesis* 4: 10–16). He also condemned the serpent to be cursed among all animals and wild creatures for tempting Eve to eat of the tree of knowledge of good and evil (*Genesis* 3:14; 2: 17). The serpent was also certainly created by the Lord God, but was "more crafty than any other wild animal that the Lord God had made" (*Genesis* 3: 1), hence his or her or its predilection to tempt Eve with his/her/its wiles. There is no biblical reference in *Genesis*, however, to other creatures, such as monsters condemned by Him, and certainly not condemned to marshes and swamps, though there is biblical reference to other watery monsters, such as Leviathan, in *Job* and the *Psalms* who may be a crocodile, the king of the jungle swamp, the uncanny creature par excellence (Giblett 2018b, chapter 4; 2019b, chapter 2).

Certainly in *Genesis* there were giants or nephilim "on the earth in those days—and also afterwards [. . .] who were heroes that were of old, warriors of old" (*Genesis* 6: 4). These giants are like Beowulf, the hero of *Beowulf,* though he does not fight with God (unlike the biblical giants), but with the monsters Grendel and his mother. Tolkien (1950, xi) remarks that Beowulf has "an addition of power, beyond the natural," in other words, supernatural; he is a super hero. By contrast, in Heaney's translation Hrothgar describes Grendel as "an unnatural birth" (*Beowulf* 1999, 45). Tolkien (1950, xi) goes on to remark that "the superhuman thirtyfold strength possessed by Beowulf (in this Christian poem [. . .] is his special gift from God)." By contrast, the two villainous monsters condemned by God are imbued with deadly demonic power. Grendel is described as "the demon, the dark death-shadow" (*Beowulf* 1950, 28; *Beowulf* 1999, 7). Whereas Clark Hall calls them "evil spirits" (*Beowulf* 1950, 25) and Heaney "evil phantoms" (*Beowulf 1*999, 6), Tolkien (2104, 16) in his recently published translation of *Beowulf* calls them "haunting shapes of hell." In his accompanying commentary Tolkien traces the etymology of the Old English words back to their associations with "dead body" and "the undead" (Tolkien 2014, 163). Whereas Clark Hall calls them "mysterious creatures of hell" (*Beowulf* 1950, 28) and Heaney (*Beowulf* 1999, 7) "reavers from hell," Tolkien (2014, 167) prefers "sorcerers of hell" and comments on hell as meaning "the underworld, Hades, the Realm of the Dead [. . .], the darkness." The swamp as hell and hell as the swamp as the unhomely home for monsters by no means begins with *Beowulf*, nor ends with it (as I have shown elsewhere; Giblett 1996).

While the story of the serpent, the temptation of Eve and the fall, and the story of Cain and Abel are all well-known, how or where the Christian basis for the story of Grendel and his mother arose or was developed or concocted out of the story of Cain, or even as monsters given birth by Cain, is another question. The reason for it lies, I think, in the shift from the sacrality of wet lands to Christianity as a religion of dry lands. One way in which Christianity asserted its dominance in England over the sacrality of swamps and marshes was figuring and condemning wetlands as hell, as a place of monsters, the basis for later draining and"reclaiming" them and enclosing the commons into private property, two processes that went hand-in-hand (as we will see in chapter 3 of the present volume). Theologically *Beowulf* is highly suspect, but evangelistically it was very effective in asserting the dominance of Christianity and ecologically it was very destructive of wetlands and their indigenous peoples and their livelihoods in asserting the dominance of dry-land agriculture.

The desirable, representative landscape of modern European Christianity is a drained wetland, first made possible by wind-powered pumps, and later in the industrial age, by steam-powered dredgers. The drained wetland represents the triumph of Christian beliefs and morality over the people of the Fens and other wetlands who valued wetlands as the place of sacrality par excellence (as we will also see in chapter 3 of the present volume and as I have shown previously; see Giblett 1996).

Wetlands were subjected to a double, double whammy in and by patriarchal society and culture. Wetlands were inimical to dryland agriculture and to the city. The hard and heavy materials of European-style cities cannot be supported in and by the soft and moist soils of wetlands. The establishment of cities in wetlands, such as London, Paris, and Venice, involved the draining and filling, or canalizing and reclamation of wetlands in the foundational event for establishing European cities, later their settler diasporas and for the age of the cities that is still ongoing. The city triumphs over the swamp, the metropolis over the marsh. The repressed wetland returns, however, in floods, as with New Orleans in the aftermath of Hurricane Katrina (see Giblett 2009, chapter 10; 2016a, chapter 10).

Wetlands were denigrated and demonized by Christianity in the foundational event for the establishment of Christian hegemony over sacrality in both Europe and its colonies. Wetlands were drained and filled by wind-powered pumps and later steam-powered dredgers in arguably the foundational event of the modern age with the development of modern industrial agriculture and cities, followed later by the expansion of suburbia into wetlands in sprawling megapolises. To add insult to injury, the slums and the dark underside of modern cities were also figured as wetlands. The repressed wetland returns in figures for slums and the dark underside of the modern city. Modern industrial technology, agriculture, cities and war created wet wastelands in arguably the foundational event of the hypermodern age (as we will see below). To add further insult to injury, the wet wastelands of World War I were also figured as wetlands, as I will show shortly with Tolkien's "Dead Marshes" in *Lord of the Rings.*

Beowulf plays a perhaps unlikely, but vital, role in this history. Regarding the Christian (and hybrid Germanic) provenance and parentage of *Beowulf,* Wrenn leaves us in no doubt when he argues in his introduction to the Clark Hall translation of *Beowulf* that the writer is:

> a Christian poet addressing a Christian audience, yet sharing with that audience a deep interest in history, legend and tradition, and presenting the inherited Germanic heroic ideals in an inherited poetical diction, [who] would not go out of his way to include anything *specifically* Christian: yet for him as for his listeners, the traditional ideals and ways of life must inevitably be "colored" by the new Christian Latin culture. (*Beowulf* 1950, 9)

Yet the Christian poet of *Beowulf did* go out of his away to include the specifically European Christian (but dubiously biblical) account of the birth and condemnation of monsters which *was* colored with a distinctly dark Germanic hue with *Beowulf*'s account of the birth of monsters and the monstrification and demonization of wetlands.

Along similar lines to Wrenn, Tolkien (2006, 22) stated in his lecture on *Beowulf* that "if the specifically Christian was suppressed [and refers in a footnote to 'the only definitely Scriptural references' to the story of Cain and Abel (Tolkien 2006, 45], so also were the old gods." This strikes me as disingenuous for despite their dubious biblical basis the specifically Christian references in *Beowulf* to monsters as Cain's kindred and to him as progenitor of monsters do work precisely and explicitly to suppress "the old gods." And goddesses. Tolkien (2006, 26) later states that in *Beowulf* "we have [. . .] an historical poem about the pagan past." Not only are its claims to historical veracity about "the pagan past" shaky as Tolkien acknowledges and as one would expect in an historical poem, but also its take on "the pagan past" is problematic given that the old gods of monsters were not stamped out by the agents of the new God and so "the pagan past," or sacrality, is still present. By contrast, Kingsnorth's *The Wake* is an historical novel about the sacral present set in the past (which I will show later in the present chapter).

In his recently published commentary on *Beowulf* accompanying his translation Tolkien (2014, 159) comments selectively on these passages and on the provenance of the monsters. He does not comment on their uncanny wetland (un)home. The setting for the poem is the mere background or backdrop against which the Christian story and heroic actions of Beowulf are set and not the context for the story, nor environment in which the story takes places, let alone an agent or actor in the plot (as is his portrayal of the Dead Marshes in *The Lord of the Rings* as we will see shortly, nor are they for Tolkien when he comes to discuss the second part of *Beowulf* in which the eponymous hero and a dragon do battle (see Giblett 2018b, chapter 3). This is one of a number of English myths of nation and empire foundation depicting the triumph of a hero over a dragon. They culminate in the legend of St George and the dragon. In some versions of this legend, the dragon is a swamp serpent or marsh monster.[1]

Yet the association between the monsters and their marshland (un)home is vital for the story (as it is for Frodo in *The Lord of the Rings*), especially as Beowulf descends, like the epic hero of Virgil, Dante, and Odysseus, into the underworld of the marsh to slay the monsters. Tolkien (2014, 16 and 158) comments at length on the anachronistic Miltonic description of Grendel as "a fiend of hell," glossed by Clark Hall as "like all other monsters, he was of the brood of Cain, and dwelt, an outcast spirit, among the moors and fens" (*Beowulf 1*950, 25). In a much more secular, but also theological key, Heaney (*Beowulf 1*999, 50) in his translation of *Beowulf* translates a later line about Grendel's mother as "that swamp-thing from hell." Precisely. I wonder if Heaney knew about the *Swamp Thing* comic series published by DC comics. The swamp as hell (and vice versa) is a persistent trope.

Tolkien (2014, 159 and 161) wonders whether "a fiend of hell" is possibly "a kind of half-theological notion" from "an unwritten chapter (as it were) of the Old Testament." Yet despite its dubious theological basis, at best half-theological and half-mythological, at worst an anti-sacral Christian concocted confection, this unwritten chapter of the Old Testament *had* to be written so that European Christianity could write the epic story of its heroic conquest of pre-Christian sacral marshes and their monsters. This chapter *was* written and the chapter is called *Beowulf.* It became a canonic text, not only of English literature, but also of English Christianity with its new creation story about marshes and monsters interpolated in *Genesis* 4:16 after Cain's punishment by God and before *Genesis* 4: 17 and Cain's creation of the city.

Cain is the father of monsters, the father of both monsters in the marsh and of the monstrous city in the marsh. The city is like a son, a bachelor birth from a bachelor machine. The city drains the marsh before it and eradicates the monster while taking up and becoming monstrous itself true to its conception. The monstrous city destroyed the monstrous marsh; the dryland city dredged and drained the wetland. Perhaps this is a case of history repeating itself, of the sin of the father of fratricide, being repeated and revisited on the next generation with one of his progeny—the city—killing another—the marsh monster and monstrous marsh in aquaterracide.

Cities built on or by wetlands are not only the work of fallen "man" in creating something that God did not create, but also of fallen "man" who rises up and wrests from God the divine function of dividing land from water (*Genesis* 1: 6), and becomes as God. The builders of the ancient city of Babel became as God by building a city that rose to the heavens; the builders of the ancient, medieval, Renaissance and modern cities set in wetlands became as God by building a city that divided land from water immersed in the earth. However, they could not become as God by creating wetlands; in fact, they destroyed wetlands in and by creating the city. Only recently has "man" become as God by creating wetlands. "Man" thus arrogated to himself the power to both destroy and create wetlands and thereby became more than God, a super God.

Beowulf as the new half-written chapter in Old English of the Old Testament would be posed in the context of the whole book of *Genesis* against *Genesis* 1: 1–2 in which God created the world as wetland, though these verses had never been read to my knowledge in this way until I first did so 25 years ago in *Postmodern wetlands* (Giblett 1996, 142–143). I also argued that *Genesis* 1: 1–2 needs to be rewritten or retranslated in pre-modern sacral and postmodern countercultural and ecological terms that recognizes the importance, even primacy, of wetlands: "in the beginning God created the heavens and the earth, and the earth was wetlands." The creation of the world as wetland was God's first and best work. When God divided the land from the water in *Genesis* 1: 9–10 it was not so much that He forgot or forsook wetlands than that He knew that the land and the water and the creatures that He later created to live in or on them would not survive and would not be able to reproduce without wetlands. Subsequent days of creation do not supersede and "forget" about their predecessors but build and rely upon them. Indeed, before God sent rain water came up from the aquifers and watered the surface of the ground. In the Judeo-Christian creation story the days have been sees in terms of a progressive time-line heading toward the ultimate goal of humanity, and not as an evolutionary process.

Without the first day of creation, the creation of the earth as wetland, there is no sixth day, no creation of humankind. Without the recreation of wetlands on the first day, and every day for that matter, there is no recreation of human life; there is no human life full stop, on the sixth or any other day. Rather than being forgotten by God, the wetland is the first day of creation, God's first and best work, without which no other day could ever begin, could ever dawn. Rather than being hell, the wetland is the closest thing to heaven because it was created next, right after heaven. When one goes to the wetland one goes back to the first day of creation when it springs fresh and new from the hand and word of God, when the spirit broods over the earth as wetland like a hovering marsh harrier, not like a malevolent ghost or swamp monster.

MILTON'S MONSTROUS MARSHES

Grendel is a forerunner to Milton's Satan as he is a swamp serpent and marsh monster, or more precisely a miasmatic marsh monster. He is described as "involved in rising mist" (IX, 75) and as "wrapped in mist/Of midnight vapour" who glides "obscure" (IX, 158–159) "like a black mist low creeping" (IX, 180). Satan is a miasmatic marsh monster because as a fallen angel he has been "mixed with bestial slime,/This essence to incarnate and imbrute" (IX, 165–6). Satan "personifies," or more precisely "monstrifies," slime and wears as his proper wardrobe the miasma which arise from it. He monstrifies slime because he was "engendered in the Pythian vale on slime" (X, 530). He is born from slime and made of slime; he is slime by birth and slime by nature. As a slimy swamp serpent his mode of propulsion is not restricted to any one kind, but varies according to the (wet)landscape over which he is travelling. As he does not stick to one mode of propulsion he is aberrant: "so eagerly the Fiend/O'er bog or steep, through strait, rough, dense, or rare,/ With head, hands, wings, or feet pursues his way,/And swims or sinks, or wades, or creeps, or flies" (II, 947–50). Satan is even a flying miasmatic marsh monster and swamp serpent. He is monstrous because he combines the physical characteristics of birds (he flies [like the Stymphalian birds]), fish (he swims), reptiles (he creeps), and quadrupeds or bipeds (he sinks or wades).

In *Paradise Lost* Milton syncretizes Greek philosophy with Judeo-Christian theology. He refers to the four elements as "the cumbrous elements, earth, flood, air, fire" (III, 715):

[. . .] ye elements, the eldest birth
Of Nature's womb, that in quaternion run
Perpetual circle, multiform, and mix

And nourish all things, let your ceaseless change
Vary to our great Maker still new praise. (V, 180–184)

The mixing of elements in "Nature's womb" does indeed nourish all things as Milton indicates. It is the wet land and warm water which give birth to life and which are the womb of the world.

Milton needs to be rewritten in environmentally friendly terms in a poem that could be called "Her Dank Materials":

Earth, air, fire and water:
The unencumbered elements,
The eldest birth of Nature's womb,
That in four ways run perpetual circle,
Multiform and mix and nourish all things,
Let your ceaseless change
Sing to our great Mater still new praise.

Rather than the womb of the world, wetlands for Milton are the tomb in the underworld. In *Paradise Lost* Milton saw the Styx, one of rivers of the classical underworld and for Dante a slimy swamp in the fifth circle of the inferno, as "the burning lake," as "abhorrèd" and as "the flood of deadly hate" (II, 576, 577). In the same book of *Paradise Lost* he elaborates later in more detail on:

[. . .] lakes, fens, bogs, dens and shades of death,
A universe of death, which God by curse
Created evil, for evil only good,
Where all life dies, death lives, and Nature breeds,
Perverse, all monstrous, all prodigious things,
Abominable, inutterable, and worse
Than fables yet have feigned, or fear conceived,
Gorgons and Hydras, and Chimeras dire. (II, 621–628)

This view of hell as swamp and swamp as hell persists through the nineteenth century and well into the twentieth century, such as in J. R. R. Tolkien's Dead Marshes in *Lord of the Rings* (as we will see shortly in the present chapter).

Milton's denigratory theology of wetlands needs rewriting in environmentally friendly terms as places of both life and death:

Lakes, fens, bogs, dens and spirits of life,
A universe of life, which God by love
Created good, only for good, not evil
Where all life dies, new life lives, and Nature breeds,

Promiscuous, all maternal, all prodigious things,
Fascinating, horrifying, and better
Than fables yet have feigned, or fear conceived,
Marsh monsters, swamp serpents and dragons fiery.

These monstrous beings embody the mixing of the four elements of earth, air, fire, and water.

Later in the same book of *Paradise Lost* Milton describes in more detail the watery world and womb of the first day of creation that morphs into Hell:

[. . .] this wild abyss,
The womb of Nature and perhaps her grave,
Of neither sea, nor shore, nor air, nor fire,
But all these in their pregnant causes mixed
Confus'dly, and which thus must ever fight,
Unless th' Almighty Maker them ordain
His dark materials to create more worlds,
Into this wild abyss the wary Fiend
Stood on the brink of Hell. (II, 910–918)

Milton's Satan, referred to here as "the wary Fiend," is a swamp serpent and marsh monster, or more precisely a miasmatic marsh monster, an elemental aberration who mixes earth, air, fire, and water.

Milton's philosophical theology of wetlands again needs rewriting in environmentally friendly terms as fertile, liminal places mixing creatively the four elements of earth, air, fire, and water:

This wild swamp,
The womb of Nature and perchance her tomb,
Of foursome sea, and shore, and air, and fire,
With all these in their pregnant elements mixed
Creatively, and which thus must ever in concord,
That the All-making Mater them ordain
Her dank materials to create more worlds,
Into this wild swamp the wary friend
Stood on the brink of Heaven.

And not on the brink of Hell for this is God's first and best work in the beginning on the first day of creation. Without it no other day could ever begin, could ever dawn. Rather than being hell, the wetland is the closest thing to heaven because it was created next, right after heaven. When one goes to the wetland one goes back to the first day of creation when it springs fresh and new from the hand and word of God, when the breath of God broods over the earth as wetland, like a hovering marsh harrier not like a malevolent ghost.

God, for Milton, brings order out of the wetland chaos, out of "the vast immeasurable abyss/Outrageous as a sea, dark, wasteful, wild" (VII, 211,2):

[. . .] on the wat'ry calm
His brooding wings the Spirit of God outspread,
And vital virtue infused, and vital warmth
Throughout the fluid mass, but downward purged
The black tartareous cold infernal dregs
Adverse to life [. . .] (VII, 234–239)

Hell as swamp was not just any part of hell for Milton, but the lowest part of the classical Greek underworld called Tartarus. Milton regards water as lifeless until the spirit of God breathes life into it. Milton's theology entails sublimation of the materiality and immanent spirituality of earth and water into the transcendental spirituality of God that has no biblical basis (see Giblett 2018a).

Milton's denigratory theology of creation again needs rewriting in sacral wetland-friendly terms:

On the wat'ry calm
Her brooding wings the breath of wind outspread,
And vital virtue suffused, and vital warmth
Throughout the fluid mess, so outward splurged
Her black, warm, slimy dank materials
Conduce for life.

A wetland-friendly philosophy and theology of creation entails desublimation of solid matter (earth) into the immanent spirituality (breath, air; warm, fiery) of liquid materiality (water).

The wetland harks back to the time when the entire earth was wetland. *Canadian Wetlands: Places and People* (Giblett 2014) begin with a rewriting of *Genesis* 1: 1–10 and 4: 16a-c:

> In the beginning there was the sky above, and the earth and water below. In the middle was the wetland. The earth and the water were without form and were chaotic. Darkness and light moved over the face and body of the earth and water. Earth and water were wet land. This was the first act of creation, the first coming into being, the first and best work of creation. The world was wetland, and the wetland was womb from which all later life sprung. Including human life made, not out of dust, into and for a desert people, but out of mud into and for marsh

> people. The wetland is the womb, and the womb is a wet land; both are a slimy swamp of embryonic life. The wetland is not only the womb but also the tomb into which all life died and from which new life was, and is, born. This was the world before the fall when the world was good, before the world fell into the knowledge of good and evil. This was world before the swamp and the marsh became the home alone of darkness, disease, death and to grotesque monsters lurking in the uncanny depths of their murky waters evoking horror and fascination in any who should be so (un)fortunate to stumble into them. (Giblett 2014, 3)

TOLKIEN'S DEAD MARSHES

Tolkien certainly buys into and reproduces the doctrine straight out of *Beowulf* that wetlands were the creation of evil after the fall and that they were not created by God before the fall. In *The Silmarillion,* the back-story and prequel to *Lord of the Rings*, as a consequence of the "evil doings" of Melkor, "green things fell sick and rotted, and rivers were choked with weeds and slime, and *fens were made*, rank and poisonous, the breeding place of flies" (Tolkien 1977, 36; my emphasis). Fens were a consequence of evil. They did not exist prior to evil. This is in line with *Beowulf* and its new chapter in Old English of the Old Testament.

Tolkien's denigration of wetlands achieves its ultimate expression in the "Dead Marshes" in *The Two Towers,* the second volume of *The Lord of the Rings* trilogy, in a chapter called "The Passage of the Marshes." Here Tolkien has his narrator describe the entry of Frodo, Sam, and Gollum into these marshes via a stream where "the gully became ever shallower and the slope of its floor more gradual. Its bottom was less stony and earthier, and slowly its sides dwindled into mere banks. It began to wind and wander" (Tolkien 2002, 630). This description of the transition from stream to marsh, of stream morphing into marsh, of a stream petering out in a marsh enacts a number of conventional geomorphological preferences, or prejudices: deep over shallow; incline over gradual slope; stony or hard bottom over soft or earthy bottom; defined sides over dwindling banks; and a straight course over a winding and wandering way.

It only gets worse:

> That night drew to its end, but clouds were now over the moon and star, and they knew of the coming day only by the slow spreading of the thin grey light. In a chill hour they came to the end of the water-course. The banks became moss-grown mounds. Over the last shelf of rotting stone the stream gurgled and fell down into a brown bog and was lost. Dry reeds hissed and rattled though they could feel no wind. (Tolkien 2002, 630)

The Dead Marshes are not only a liminal place between land and water but also a twilit crepuscular place between night and day, dark and light. The coursing water ends and stops flowing. The dwindling banks morph into moss-grown mounds. The stream stops abruptly and is lost in a brown bog. The green world is lost in brown wetland or wasteland for Tolkien. Despite being a wetland, dry reeds hiss like snakes or serpents.

The Marshes continue to get worse when "on either side and in front wide fens and mires now lay, stretching away [. . .] into dim half-light. Mists curled and smoked from dark and noisome pools. The reek of them hung stifling in the still air" (Tolkien 2002, 630). The initial sensory encounter entering the Dead Marshes is via the distancing sense of sight whereas this more immediate experience is via the up-close sense of smell. Typically the marsh is a place of miasma and effluvia (Giblett 1996).

And so it goes on:

> It was already day, a windless and sullen morning, and the marsh-reeks lay in heavy banks. No sun pierced the low clouded sky . . . They were soon lost in a shadowy silent world, cut off from all view of the lands about . . . What had looked liked one vast fen was really an endless network of pools, and soft mires, and winding half-strangled watercourses. (Tolkien 2002, 632)

Explorers often complained about getting lost in wetlands because they obstructed the view.

To reach the nadir where "it was dreary and wearisome. Cold clammy water still held sway in the forsaken country. The only green was the scum of livid weed on the dark greasy surface of the sullen water. Dead grasses and rotting weeds loomed up in the mists like ragged shadows of long-forgotten summers" (Tolkien 2002, 632). The marsh becomes "it," a thing, a swamp thing, and ceases to be a living being. It is also melancholic. It not only assaults and offends the sense of smell but also affects adversely the psychology of the marsh traveller. It is figured as a place of death, whereas it is a place of both life and death, life decomposing, dying and regenerating into new life. I am reminded of the souls of the sullen stuck in slime in Dante's *Inferno* (we will see in the following chapter of the present volume).

Yet there is always a beacon of Christian light in this benighted place:

> As the day wore on the light increased a little, and the mists lifted, growing thinner and more transparent. Far above the rot and vapours of the world the Sun was riding high and golden now in a serene country with floors of dazzling foam, but only a passing ghost of her could they see below, bleared, pale, giving now colour and no warmth [. . .] There was a deep silence, only scraped on

> its surfaces by the faint quiver of empty seed-plumes, and broken grass-blades trembling in small air-movements that they could not feel. (Tolkien 2002, 632)

The narrator takes a privileged, God-like position above the quaking zone of the wetland scene to describe the sun shining on all below her. The sun is feminized in line with the female Maia Arien who drives the chariot of the Sun in *The Silmarillion*, as opposed to the male Maia Tillion who drives the chariot of the Moon. This "gendering" of the Sun and Moon is already there in the earliest version of Tolkien's mythology in *The Book of Lost Tales* composed in the 1910s–1920s. What is the marsh? How is the marsh gendered? Typically and traditionally it has been feminized.

The marsh is a place not only of melancholy, but also of mourning: "Not a bird!" said Sam mournfully. "No, no birds," said Gollum [. . .] "No birds here. There are snakeses, wormses, things in the pools. Lots of things, lots of nasty things. No birds" (Tolkien 2002, 632). The marsh is a place of swarming lower creatures in the great Christian hierarchical chain of being.

The wetland becomes even wetter as "the fens grew more wet, opening into wide stagnant meres, among which it grew more and more difficult to find the firmer places where feet could tread without sinking into gurgling mud [. . .] Presently it grew altogether dark: the air itself seemed black and heavy to breathe" (Tolkien 2002, 633).

Garth (2006, 46) comments on this and other passages in *The Lord of the Rings* about "the very air of the nightmarish Dead Marshes" that "there are grounds to suspect that Tolkien was influenced by his experience [as a soldier in World War I] of poison gas as he devised a symbolic shape for battlefield trauma, demoralization and despair." Yet the heavy air and marsh gases of wetlands is not the result of poison gases produced by modern industry and modern warfare, but of decomposing organic matter.

Tolkien does not distinguish (and Garth follows suit) between a wetland as a native quaking zone made by ancestral beings and traditional human beings, on the one hand, and on the other, a wet wasteland as a feral quaking zone made by modern industrial hands, such as in the case of Somme mud of the World War I that he experienced personally. Tolkien (1981, 303) wrote in a letter that he did not think "either [World] War [I or II...] war [. . .] had any influence upon either the plot or the manner of its [*The Lord of the Rings*] unfolding. Perhaps in landscape. The Dead Marshes and the approaches to the Morannon owe something to Northern France after the Battle of the Somme." After citing this extract, Garth (2003, 310) goes on to relate that for Hugh Cecil in his survey of British fiction of World War I the Dead Marshes is "a scene of morbid desolation that has become in effect, a shorthand symbol for the trenches" (Garth 2003, 311). The Dead Marshes were not inspired by a living, native wetland, but rather by a dead, feral wet wasteland, a counterintuitively artificial wetland (see Giblett 2009, chapter 1). The Dead Marshes then became a shorthand symbol for the scene of morbid desolation of the feral wet wasteland of the mud of the Somme created by modern industrial warfare and its technologies (see Giblett 2009, chapter 4). Unlike the Somme in World War I, the Dead Marshes were not created by war, but like the Somme they were a burial ground for the war dead.

What could be called a circularity of figurality occurs here in what I call placism: a recently human-made wet wasteland, such as Somme mud, is made to figure a native or ancient wetland, such as the marsh that the Dead Marshes once were, and the pejorative associations of the former are applied to the latter. Both are places of death and the dead, whereas marshes are places of both life and death. The obverse also applies where the pejorative associations of a swamp are applied to the city slum and consequently the swamp comes to be seen in the same pejorative light (or darkness) as the dark underside of the city in its slums (Giblett 2019b, d). Both are places of death and the dead, whereas marshes are places of both life and death. This pejorative figuring of wetlands becomes mutually reinforcing and culturally naturalized, but its logic is circular and its ideology is placist. Tolkien and other placist writers (such as many nineteenth-century writers about the dark underside of the city) not only ascribe moral qualities to a place undeserving of such opprobrium, but also give it derogatory qualities in the lower psychopathological register (such as those nineteenth-century writers about the dark underside of the city who figure it as stagnant pool, bottomless abyss, nether world, or dreadful night; see Giblett 2019b, chapter 3).

The history of the Dead Marshes bears out the conclusion that they were created as a burial ground for the dead. The Tolkien Gateway (n.d.) says that:

> In the year S.A. 3434 the host of the Last Alliance of Elves and Men fought the forces of Mordor in the Battle of Dagorlad. During the battle on the plains more than half of the Elves of Lothlórien under the command of King Amdír were driven into the Dead Marshes. After the battle many of the slain were buried outside of the marshy area but over time (in the Third Age) the Marshes had grown and swallowed the graves. In T.A. 1944 King Ondoher of Gondor was caught by a surprise attack by the Wainriders upon the Dagorlad. When the King and his guard were destroyed, many of the confused soldiers of Gondor were driven by the attackers into the Dead Marshes. However, after Eärnil II won the Battle of the Camp, those of the Wainriders who were not slain in the fight were themselves driven into the Dead Marshes and there perished.

Frodo, Gollum and Sam have entered the quaking zone where the earth trembles. They have also entered the dead zone of "dark water," "slime," "sticky ooze" and the marsh lights of burning marsh gases described by

Gollum as "candles of corpses" for this is a place of the dead with "dead faces in the water" (Tolkien 2002, 633). "Fire on the flood" has a direct prototype in *Beowulf* (*Beowulf* 1950, 90) for the burning marsh gases in *Two Towers*. Tolkien survived Somme mud just as Frodo, Sam and Gollum at last:

> Came to the end of the black mere, and they crossed it perilously, crawling or hopping from one treacherous island tussock to another. Often they floundered, stepping or falling hands-first into waters as noisome as a cesspool, till they were slimed and fouled almost up to their necks and stank in one another's nostrils. It was late in the night when at length they reached firmer ground again [. . .] [Gollum] in some mysterious way, by some blended sense of feel, and smell, and uncanny memory for shapes in the dark, [. . .] seemed to know just he was again, to be sure of his road ahead. (Tolkien 2002, 634–635)

The human-made cesspool is made to figure the smell of the natural marsh, albeit in this case a burial ground. In the Dead Marshes Gollum did not know where he was and he was unsure of the road ahead for there was no road. The Dead Marshes are a trackless and roadless waste in which it is easy to get lost. Gollum does not rely on the aesthetic senses of sight and hearing to find his way again but on the uncanny senses of smell and touch to feel and intuit his way.

KINGSNORTH'S *THE WAKE*

Published in 2014, the same year as Tolkien's translation of, and commentary on, *Beowulf*, and written in what Kingsnorth calls, and devised as, "shadow English," "a pseudo-language" that might have been that adumbrates Old and Modern English, *The Wake* is an historical novel set in the aftermath of the Norman invasion of England in 1066, and geographically in and around the Fens. Unlike Tolkien for whom the marsh is the mere backdrop for human action, or the home of horrific monsters, or a horrific place itself, for Buccmaster, the first-person narrator of *The Wake*, "the fenn" "was what has macd us who we is as folc this place" (Kingsnorth 2014, 33). The shallow waters of the wetland are fascinating and delightful as they are referred to as "the deorc undeop waters of the fenn" (34). The dark waters of the wetland are not an object, or "abject" to use Kristeva's (1982) more apt term, of horror that hides lurking monsters unlike for the *Beowulf* poet and Tolkien. Buccmaster acknowledges that "the fenn that can be so blaec and deop and cold on this mergen was a thing of great wundor" (50). Buccmaster has heard the Beowulf story. When he is a child he goes out in his grandfather's boat into the fen and fears that the world under the water is where Grendel and his mother were living and that they "was cuman for me" (51). The world under the water is the home, not to monsters, but to the old sacrality and folk knowledges. It is also the site of resistance to the Normans as it was for the historical Hereward of the same period usually known as "the wake," but Kingsnorth seems to be implying or suggesting that Buccmaster is another wake, or the real wake.

The fens are not necessarily a pleasant place but they are subject to commonplace misconceptions. Buccmaster relates "it is saed that the fenns is all deorc waters and deope mud" and "that folcs of the fenns" is "yfel wights of the mere." The fens are not the place of evil monsters. Buccmaster counters these misconceptions:

> These is the things folcs saes who does not cnawan naht of what they is specan for the fenns is a place of wundor to those who cnawan them. yes there is deop meres and waters so blaec that oxen is lost in them and nefer seen and there is muds what strecces for miles by secg and lesch [reeds] . . . there is the blaec fenns where the eorth and the waters in all blaec like the graef. (Kingsnorth 2014, 123)

Buccmaster acknowledges the unpleasant aspects of the fens, including their black water.

Buccmaster initially seeks shelter from the Normans in the wood but later comes to the realization that "I was a man of the water not the wud the fenns was callan me" and that he must go to "the water lands for there was no other place triewe in this world" (Kingsnorth 2014, 286). When he returns to the fens it is a return to his home land with familiar black earth, meres and sedges and "in the heofon was the eald feel of my life and of my folc" (287). He acknowledges that "i missed the fenns" and that "it is the triewest place in angland" (287). The fens are the quintessential English home land, the repository of sacrality (as we will see in chapter 3 of the present volume). Long may they live.

NOTE

1. For environmentally friendly retellings and ecocritical readings of the story of Beowulf and the dragon doing battle in the second part of *Beowulf*, and of the legend of St George killing the dragon as portrayed in books, statues, names of streets, buildings and a bank, and so on, and on coins, see Giblett (2018a, c, 2019a, c).

Chapter 2

Psychology of Wetlands

Mourning, Melancholy, and Marshes

Wetlands were once thought to be bad for the body. According to the miasmatic theory of disease, malaria—literally "bad air"—could be caught be breathing miasma or bad air. Wetlands could also be bad for the mind. Indeed, they can plunge the mind into melancholia, and even into madness and despair. This chapter begins by considering John Bunyan's "slough of despond." Despondency is another name for melancholy that is the immanent counterpart to the transcendental sublime and the spiritual counterpart to the psychological uncanny. As the uncanny is a secular theology for a world in which God is dead, so melancholy is a secular spirituality in the lower psychopathological register of what Julia Kristeva (1982) calls the abject, the mediating category between subject and object that makes both possible.

Bunyan and detective fiction associate the slough of despond with the Psalmist's "valley of the shadow of death." The detective descends into the alley of the shadow of death in the infernal slums of the modern city, a secularised slough of despond. Melancholy is associated with death, or at least with the (v)alley of the shadow of death (the precursor to death) and with mourning (the aftermath of death). Melancholia and mourning are connected in Sigmund Freud's essay on the topic, in Walter Benjamin's work on the German mourning play, in E. M. Cioran's discussions of despair and melancholy, in Charles Dickens' depiction of "the grim domains of Giant Despair" and in Vera Brittain's account of her visit to the swampy city of Venice in the aftermath of muddy World War I as the present chapter goes on to discuss.

In a typical assessment made more than one hundred years ago the later-to-be president of the United States Theodore Roosevelt (cited by Brooks 2014, 111) referred to "the melancholy marshes" in the "lonely lands" of "the wilderness." More specifically, Hubert Davis (1962, 21) referred to "the stench of a rotting carcass of some depraved hunter lost in the depths of its [the Great

Dismal Swamp's] melancholy jungles."[1] The corpse beneath the swamp is not only a reminder of death, but also a vector of melancholy. This assessment of the marsh as melancholic can also be found in the work of those masters of dejection, the romantic poets. Percy Bysshe Shelley (1975, 51–52), for example, referred to "a wide and melancholy waste/of putrid marshes" when evoking "the Spirit of Solitude." Previously Samuel Pepys (1972, 311) in the late seventeenth century travelled "over most sad Fenns (all the way observing the sad life that the people of that place . . . do live)." More recently, the map of "the world" of those 1920s classics of Anglophone children's literature, *Winnie-the-Pooh* and *The House at Pooh Corner*, with its Christopher Robin of the clean and proper body, has that arch manic-depressive donkey Eeyore living in "Eeyore's Gloomy Place" which is "rather boggy and sad" (Milne 1958, frontispiece map). Other examples in adolescent literature are the Swamps of Sadness in Michael Ende's *The Neverending Story* and the Dead Marshes in J. R. R. Tolkien's *The Lord of the Rings* (as we saw in the previous chapter).

THE SLOUGH OF DESPOND

The most famous depiction in English culture of a depressed place (in two senses) is John Bunyan's "Slough of Despond" in *Pilgrim's Progress,* first published in 1678. Louis Marin (cited by Fritzell 1978, 528) has described *Pilgrim's Progress* as "the most influential religious book ever composed in the English language." It is also the second most published book in the English language after the Bible and before the complete works of Shakespeare. No doubt part of its preeminent influence has been to educate generations of readers that an ecologically functioning wetland is not what it seems, and is, but an allegorical emblem for a sump of iniquity that would drag the unsuspecting and unwary Christian down and entrap him/her for all eternity. Bunyan (2008, 18) relates how "this miry slough is such a place as cannot be mended; it is the descent whither the scum and filth that attends conviction for sin doth continually run, and therefore it is called the Slough of Despond." Bunyan's theologised slough of despond is singing from the song sheet of the melancholic swamp composed by Robert Burton in his miasmatic *Anatomy of Melancholy* published in 1621 (Burton 1932, 224–225, 239–240; see Giblett 1996, 161–162).

The same figure is used with the same denigratory overtones (though not of course as an allegorical emblem for sin) in Donna Haraway's bald proclamation that "like Christian in *Pilgrim's Progress* [. . .] I am committed to skirting the slough of despond and the parasite-infested swamps of nowhere to reach more salubrious environs" (Haraway 1992, 295; see 329, n.1 for her exclusion of the allegory of sin as slough of despond in invoking *Pilgrim's Progress*). One wonders salubrious for whom? In whose terms? And if the slough of despond is not an allegorical emblem for sin, what is "sin" in Haraway's theologised, or at least moralised, wetlandscape? What is "nowhere"? Is it "the amniotic effluvia of terminal industrialism" that Haraway refers to later (but without making any explicit connection) and that we have encountered earlier? If so, then the figure would have a critical edge and contemporary pertinence; if not, then Haraway's use of the figure would seem to be gratuitously uncritical about the use of tropes and blind to the politics of place (place-blind and even misaquaterrist and placist).

Yet she argues in these very same two pages that "nature is [. . .] a *topos*, a place," that "nature is also *trópos*, a trope" and later for a "politics of articulation" which speaks with an intersubject rather than for "a politics of representation" which speaks for and on behalf of an object (Haraway 1992, 296 and 311–313). But as nature is also place*s, topoi*, as different *topoi* are troped in different ways with some being valorised at the expense or to the detriment of others, so there is a politics of the articulation of tropes of topes, a politics of *trópoi* of *topoi*, not least of wetlands. Haraway cares about "the survival of jaguars and the chimpanzee, and the Hawaiian land snails, and the spotted owl, and a lot of other earth*lings*" (Haraway 1992, 311; my emphasis), but does she care about the survival of sloughs (of despond), (parasite-infested) swamps, and uncharismatic micro-fauna (like parasites)?

No such qualms of conscience or ethical dilemmas were to trouble English writers after Bunyan despite the increasing secularization of English culture. The slough of despond was divested of its religious overtones, but not of its denigratory, misaquaterrist associations. Instead of being a place of evil construed in religious terms, it became a place of melancholia, a kind of secular despondency leading to despair. Charles Dickens, for example, has "the Slough of Despond" in *Hard Times*, as well as "the slough of inanity." "The social swamp" as Thomas Huxley (1989, 335) called "*la misère*" in 1888 could be "a Slough of Despond." Dickens saves his most scathing attack on the swamp as "the grim domains of Giant Despair" and melancholy for chapter 23 of *Martin Chuzzlewit* (as we will see in the present chapter).

The slough of despond, rather than being rehabilitated or its denigratory connotations reversed in the process of increasing secularization, came to stand for the dark side of the Enlightenment and Romanticism despite their differences. The Slough of Despond, Peter Fritzell has argued:

> is the well-remembered antithesis to enlightenment and romantic thought [. . .] With the declaration of man's [*sic*] perfectibility and the affirmation of sublime and picturesque vistas, with the denial of man's imperfectibility and the negation of landscapes which do not fit the conventions of the sublime or the

> picturesque people of the nineteenth century can proceed to their new-found manifest destiny. They can proceed with the story of exploitation, the story of draining, ditching, clearing, and filling, which will improve, civilise, humanise, and finally redeem the nonhuman environment. (Fritzell 1978, 529)

Such a place as the slough of despond as the antithesis of enlightenment would seem to make the ideal setting or backdrop for the private detective story as it is a place of almost impenetrable evil and darkness which can only be pierced by the penetrating light of reason brought to bear upon it by the superior intellect and insight of the great man himself (for it is invariably a man so why not a feminist detective story which inverts, or better still subverts, this scenario?). It is hardly surprising then that the most famous private detective of them all in probably his most famous story of all should have to deal with a slough of despond. This is, of course, Sherlock Holmes in *The Hound of the Baskervilles* with its great Grimpen Mire (see Giblett 2018a, 105–106).

Similarly, Fergus Hume in his even better-selling detective novel than *The Hound* and "the first popular Melbourne novel" according to John Arnold (1983, 7–8), *The Mystery of a Hansom Cab*, first published in 1886, evokes vividly the Dantesque "Infernal Regions" "off Little Bourke St" in the alley of the shadow of death. Like the epic hero descending into the underworld to slay monsters and return home triumphant, "the detective led the way down a dark lane, which felt like a furnace owing to the heat of the night" (Hume 1999, 178–179). The infernal distress of the slums of Bourke Street can be contrasted with what Marjorie Clark (in Arnold 1983, 69) in 1927 described as "the cool charm of its [Collins Street's] arcades." Worse for the detective entering the slums of Bourke Street, "it was like walking in the valley of the shadow of death [. . .] And, indeed, it was not unlike the description in Bunyan's famous allegory what with the semidarkness, the wild lights and shadows" (Hume 1999, 180). Hume combines Dante and Bunyan (and the biblical Psalmist, as Bunyan is drawing on *Psalm* 23) in an omnibus and ominous grab bag of tropes for the urban underworld.

As an aside here, these pages are excerpted and reproduced as a quintessential depiction of the "mean streets and back alleys" of late nineteenth-century Melbourne as depicted in crime fiction for the anthology *Literary Melbourne* edited by Stephen Grimwade (2009, 187–190). The reference to "the valley of the shadow of death" is later misquoted as "the shadow of the valley of death" (Grimwade 2009, 244). This misquotation gives a new nuance to the idea of Death Valley transposed here from the remote deserts of California to inner-city Melbourne. Both are hot and deadly places. As Michael Cannon (1976, 42 and 43) puts it, "the Angel of Death came early and stayed late in the Melbourne of 1892 and 1893" with "epidemics of influenza, typhoid and measles [. . .] which [. . .] killed thousands." It is only a hop, skip, and a jump from Bunyan's and the Psalmist's "valley of the shadow of death" to Hume's alley of the shadow of death and Grimwade's shadow of the valley of death.

The detective's journey into the urban underworld is not only physical but also moral and allegorical. The detective novel is a secular allegory of literal and spiritual descent and ascent, of degradation and salvation, just like the sacred allegory of Bunyan's *Pilgrims Progress*. In the case of Sherlock Holmes, the detective's journey also includes a descent into the Bunyanesque "Slough of Despond" of great Grimpen Mire in *The Hound of the Baskervilles* (as we have already seen). Similarly, Flinders Street in 1845 was described as "that slough of despond" (cited by Annear 2014, 40). The swamp outside the city in the country, or inside the city in the muddy streets or slums, is a secularised satanic space. Perhaps it is fitting that the *flâneur* is a creature of the arcades, which Johann Geist (1983, vii) describes in his monumental history of the arcade as "a secularised sacred space." The swamp is a secularised satanic space for the detective that is a sacred space and place for indigenous, traditional peoples. The secularised satanic space of the swamp could be the native or natural swamp in the country, or the feral or cultural "swamp" of the slums in the city in which the latter is used as a figure for the latter.

In *Pilgrim's Progress* Bunyan (2008, 65) takes the fourth verse of *Psalm* 23 referring to the valley of the shadow of death for his text and sermonises on how "Christian must needs go through it, because the way to the Celestial City lay through the midst of it." Typically the detective in the modern city must go through the (v)alley of the shadow of death of the underworld of the slums in order to protect the upper city of the upper ten thousand living in the upper world of the crystalline celestial city. Or the hero of the modern epic of the novel, such as Jean Valjean in Hugo's *Les Miserables*, must descend into the underworld of the sewers of Paris (see Giblett 2016a, chapter 3). The modern hero of the tourist can even follow in Valjean's footsteps and go on a tour of the sewers of Paris in the marshy underworld of "Lutetia," the Latin-cum-Celtic name for the "filthy marsh" in which the city was founded. Although Melbourne as the "Paris of the South" does not have such an official tour of its sewers, "a small group has for several decades mounted regular unlawful explorations of Melbourne's stormwater drains and tunnels" in what Kristin Otto (2005, 175) calls "a Yarra [River] underworld." Sophie Cunningham (2011, 147) also comments that "while Melbourne does not boast Paris's hundreds of kilometres of underground tunnels, it does have Anzac, a cavernous drain under South Yarra where many parties have been held over the years."

Later in *Pilgrim's Progress* Christian is informed that "the valley [of the shadow of death] itself" is:

> as dark as pitch: we also saw there the hobgoblins, satyrs, and dragons of the pit; we heard also in that valley a continual howling and yelling, as of a people under unutterable misery, who there sat bound in affliction and irons; and over that hung the discouraging clouds of confusion; Death also does always spread his wings over it. In a word, it is every whit dreadful, being utterly without order. (Bunyan 2008, 67 and 69)

As the underworld is a monstrous place itself inhabited by monsters, such as dragons, so is "the monster city." In similar vein, Little Bourke Street in *The Mystery of a Hansom Cab* has its "weird and grotesquely horrible" inhabitants (Hume 1999, 220). In a word, they are monstrous. Little Bourke Street contrasts with Bourke Street proper, not only in the type of inhabitants, but also in the mode of illumination. Bourke Street is described "as the brilliantly lit street" produced by "electric lights," which highlight the members of the crowd caught in "the full glare of the electric light." By contrast, Little Bourke Street is lit by "sparsely scattered gas lamps with their 'dim light,' whereas the lanes have no lights or lamps so they are dark, or 'not quite dark,' for the atmosphere had that luminous kind of haze so observable in Australian twilights, and the weird light was just sufficient to make the darkness visible" (Hume 1999, 178–180). Typically the fictive detective of the modern city enlightens the benighted, brings light to darkness and illuminates the crepuscular gloom of crime and grime.

In her biographical and critical study of Hume's novel, Lucy Sussex (2015, 7) argues that it had "an important role in establishing detective fiction as a publishing category" as Hume is "one of the most influential crime writers of all time" who wrote "the biggest-selling crime novel of the nineteenth century, and one of the most important Australian books ever." Sussex traces the etymology of the word "detective" and how it literally means "de-roofing" as "a detective raises the roof, figuratively." The detective in the detective story, and the detective storyteller, raises the roof of dwelling spaces, looks inside and reveals what is inside to the reader. The detective and detective story reader are positioned as snooping voyeurs.

Sussex (2015, 9–10) also traces how this uncovering and de-roofing had a demonic function and cites Dickens in *Dombey and Son* published in 1848 who pleaded for "a good spirit who would take the house-tops off." David Grann (2017, 57) similarly traces how "the term 'to detect' derived from the Latin verb to 'unroof,' and because the devil, according to legend, allowed his henchmen to peer voyeuristically into houses by removing their roofs, detectives were known as 'the devil's disciples.'" In the Sherlock Holmes' story "A Case of Identity," Conan Doyle has Holmes describe the detective's work of unroofing in similar terms (albeit without the etymology and the theological overtones). The detective in the detective story, and the detective storyteller too, perform these socially useful, but morally ambiguous, roles of the good, or demonic, spirit. Both were *flâneurs* who entered "the city's central hell [of] the slums," as Sussex (2015, 60) puts it in relation to Hume and *Hansom Cab*. As with the circles of Dante's hell, or inferno, the city has a centre, and its centre is the lower depth of its slums and sewers, the grotesque lower urban strata. The seemingly most secularized literary genre of the detective story about the modern city has strong theological undertones.

MOURNING, MELANCHOLY, AND FREUD

Despondency is another name for melancholy that is the immanent counterpart to the transcendental sublime and the spiritual counterpart to the psychological uncanny. As the uncanny is a secular theology for a world in which God is dead, so melancholy is a secular spirituality in the lower psychopathological register of the abject, Kristeva's (1982) term for the mediating category between subject and object that makes both possible. In 1933 Gershom Scholem (Benjamin and Scholem 1992, 81) included a poem in a letter to Walter Benjamin the last line of which concludes that "where God once stood now stands: Melancholy." Melancholy is the abject spirituality of a world in which God is dead. Melancholy is also, as Terry Eagleton (1986, 41) argues, "the appropriate neurosis for a profit-based society." Capitalist society is driven by greed for, and indulges in gluttony of, the earth's resources, one of the seven deadly sins in the Christian tradition.[2] Just as the aesthetic and philosophical sublime lifts one up to the heady heights of intellection and theory close to the divine, so the psychological and spiritual melancholy of the uncanny depresses one down into the grotesque lower bodily and earthly strata of slime close to the demonic beneath what Gérard de Nerval calls "the black sun of melancholia" (see Kristeva 1989, and figure 2.1; see also Giblett 1996, chapters 2 and 7; 2013, chapter 14):

Melancholy is associated with death, or at least with the valley of the shadow of death (the precursor to death) and with mourning (the aftermath of death). Melancholia and mourning are connected for Freud. In his 1917 essay, "Mourning and Melancholia," Freud argued that mourning is "commonly the reaction to the loss of a beloved person," whereas melancholia "may be a reaction to the loss of a beloved object" (Freud 2005, 203 and 205). The loved object which is lost for the melancholic is specifically the breast of the mother and the water which is breast milk or the water of the wetland, the first water which nourished life (on earth), the breast of mother earth.

More broadly, the loved and lost object can be "nature" generally as Freud (2005, 195–200) identified and discussed in his essay "On Transience" published in 1916, the year before "Mourning and Melancholia." For Freud (2005,

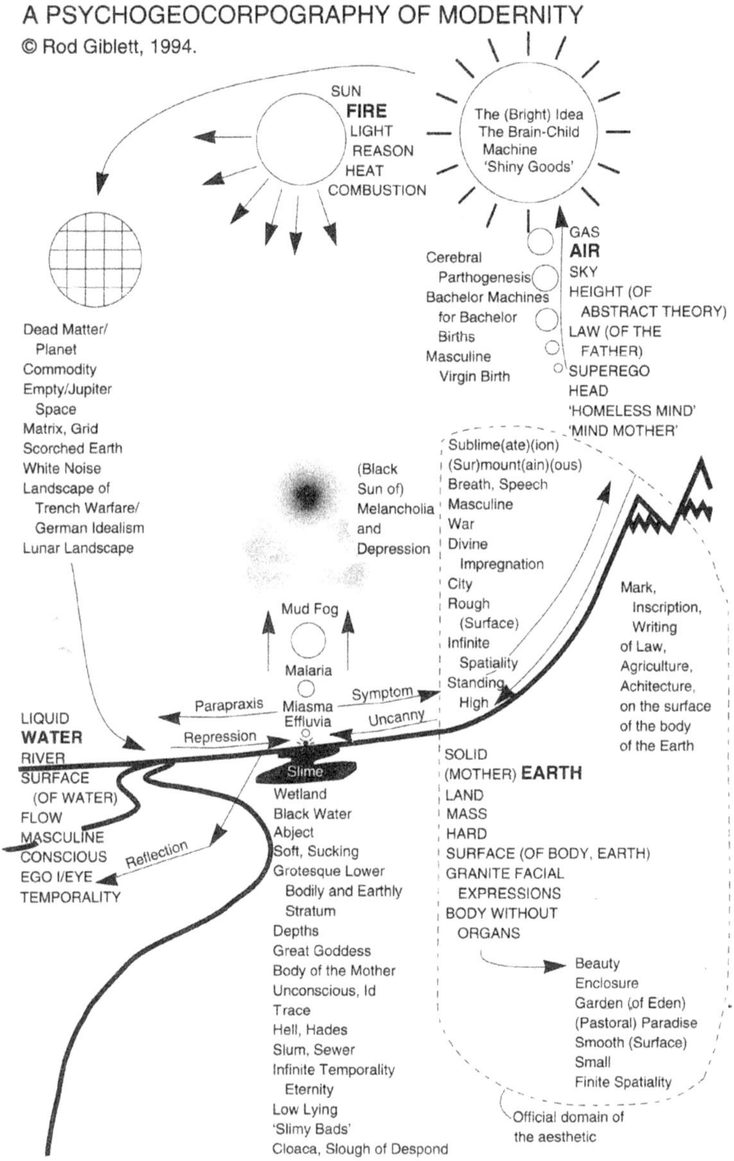

Figure 2.1 A Psychogeocorpography of Modernity by Rod Giblett. *Source*: © Rod Giblett 1994. First published in Rod Giblett, *Postmodern Wetlands: Culture, History, Ecology*, Edinburgh University Press, 1996.

199), mourning and melancholia were particularly acute during the destruction of World War I that "robbed the world of its beauties" and "destroyed [. . .] the beauty of its landscapes." Freud (2005, 198) found some consolation in the cycle of the seasons in which "the beauty of nature [. . .] returns the following year after the ravages of winter" and in the fact that "that return may be seen as eternal in terms of the length of our lives." The melancholic ego, however, is immersed in the moment and does not have the time-scale of the eternal. Freud (2005, 197) describes a young poet who demonstrates the affliction of, and fixation on, transience that is the topic of the essay and the impetus for writing it. The young poet values transient objects because of their "scarcity over time" and so experiences the "painful world-weariness" of melancholia. By contrast, Freud values enduring objects because of their generosity over time and so experiences the pleasurable world-liveliness of eternality.

Instead of investing desire in the object of love, such as the breast and/or the mother and gaining some return of pleasure on that investment (the economic metaphor is appropriate), the melancholic ego "wants to incorporate this object into itself, and in accordance with the oral or cannibalistic phase of libidinal development in which it is, it wants to do so by devouring it," or by simply "eating it" in the less orally sadistic terms of the more recent translation (Freud 1984, 258; 2005, 210). In environmental terms, the melancholic wants to incorporate the nourishing qualities of the living waters of the wetland breast into himself by devouring it through drainage or filling, and even by creating artificial ones. The object of investment was initially an object of love which was later lost.

Freud goes on to distinguish between mourning and melancholia by arguing that "in mourning it is the world which has becomes poor and empty; in melancholia it is the ego itself" (Freud 1984, 254; 2005, 205–206). In mourning the world is experienced as loss, whereas in melancholia the ego is experienced as lost. Freud outlined the process whereby melancholia establishes an "identification of the ego with the abandoned object" (Freud 1984, 258; 2005, 209): "the shadow of the object fell upon the ego [. . .] as though it were an object, the forsaken [or "abandoned"] object. In this way the loss of the object was transferred into a loss of ego" (Freud 1984, 258; 2005, 209). In melancholy, as Mikkel Borch-Jacobson (1989, 183) concludes in his discussion, "the ego becomes the object," or even more succinctly, "the lost object [is] me" as Nicolas Abraham and Maria Torok (1984) put it, as a way of denying or disavowing the loss of the loved object in narcissistic inversion.[3]

MELANCHOLY, DESPAIR, AND CIORAN

Melancholy differs from despair. In melancholy the ego is an object of ambivalent love/hate for the subject, whereas in despair the ego becomes an object of enmity and hatred for the subject. Despair in these terms is put no more eloquently and succinctly than in Shakespeare's *Richard III* (2: 2, 36–37) when the widowed Queen (whose husband, King Edward IV, was murdered) says, "I'll join with black despair against my soul. And to myself become an enemy." E. M. Cioran (2010, 3), the doyen of dark despair, uses these two lines from *Richard III* as the epigraph to the "Directions for Decomposition," the first part of *A Short History of Decay*. In despair, the lost object of the ego is the enemy, an object of hate for the subject, whereas in melancholy the ego and the lost object are one and the same in ambivalent self-love/hate. Melancholy and despair also differ in colour-coding in mainstream western culture. For Cioran (1995, 43), also the master muse of modern melancholy, "blue is a soothing color for melancholy,"[4] whereas black is the distressing color of despair.

For Cioran (1995, 15) despair, "more than any other feeling, establishes a correspondence between our being and the environment. In fact, despair requires a corresponding environment to such an extent that if needs be, it creates it." Despair festers in the feral quaking zone of the dead black waters of industrial wastelands. Confronted with the horror of the filthy, smelly mud on the Western Front in World War I produced by the terror of artillery bombardment dispensed by the monstrous machines of industrialized warfare, the self/land is made to hate itself and war is fought against the self/land. The wastelands of the cities reduced to rubble and ruins in World War II produced by the terror of firebombing dispensed also by the monstrous machines of industrialized warfare followed suit and fitted the bill too.[5] More recently, in what Picon (2000) calls "the anxious landscape" of "regions of ruin and rust" in the steel belts of post-industrial wastelands, industrial technology and infrastructure decay and decompose in pools of stagnant, and possibly toxic, water.

Like despair, melancholy also establishes a correspondence between human beings and "the environment" (natural or cultural), requires a corresponding environment and, if needs be, creates it (in wetland tropes) with "the slough of despond" and with what Cioran (2010, 110) figures as "languid mud" in a section "On Melancholy" of *A Short History of Decay*. Melancholy for Cioran (2010, 110; his emphasis) is "the *dream state of egoism*: no longer any object outside oneself, no reasons for hate or love, but that same fall into a languid mud, that same circling of the damned without a hell." By contrast, in the blackness of despair oneself is the object of hate inside oneself (as the widowed Queen puts it in *Richard III*). Languid mud for Cioran is the poetic dreamscape of melancholy and the dream space necessary for it.

Cioran also associates the swampy city of Venice with muddy melancholy. For Cioran (1995, 92), "Venice is not a historic reality but a function of melancholy, a town of tears caught between doubts and dreams." It is also caught between dread and death, as Vera Brittain demonstrates on both counts (as we will see later in the present chapter). Melancholy requires a corresponding environment and, if needs be, creates it, not only in languid mud, but also in wetland cities, such as Venice, caught between the dreams of draining and the doubts of drowning. Melancholy produces the draining dream of the eternal and serene city (as with Venice), as well as the drowning nightmare of the decaying and decomposing city that was built in wetlands. In the case of Venice, "the *alta aqua*" of floods caused increasingly by rising sea levels in the age of global heating brings the historic reality of the muddy, melancholic city back to the surface from the swamp from where it came in the return of the repressed.[6]

Cioran figures melancholy not only as languid mud functioning in the swampy city of Venice, but also in miasmatic and malignant tropes in a secular hell. Melancholy for Cioran (2010, 110) "requires a debauch of space, an infinite landscape in order to spread out its sullen and vaporous grace, its shapeless evil." Cioran is secularising and seems to be paraphrasing Dante's description of the fifth circle of hell with the souls of the sullen stuck in slime in "a dreary swampland, vaporous and malignant" (as Dante puts it in canto 7 of his *Inferno*). The secular modern melancholic is damned for the time being to, and within, a swampy hell figured in denigratory mainstream terms; they have a circle of hell and an environment in it. Secularised modern melancholy requires and, if needs be, creates the infinite landscape of the figuratively debauched and morally and aesthetically ambiguous swamp of Cioran's dreamscape. Without this dreamscape, without this surface or screen to project its dream state, or nightmare, of egoism onto, melancholic death in life and ambivalent self-love/hate would not have a place to breed the self-hatred of black despair that can lead to the end of life in self-death.[7]

The languid mud of melancholy mixes the elements of earth and water; melancholy is muddy so it flows slowly; melancholy is stagnant. Melancholy for E. M. Cioran (2013, 85; his emphasis) is not only "languid mud," but also "*stagnant* blood." Whether it is languid brown mud, billowing black bile, stagnating red blood, or slimy white phlegm, melancholy, like water in and through wetlands, flows slowly. In the theory of the humors, melancholy is associated with bile, phlegm with the phlegmatic, and blood with the sanguine. They all mediate solid and liquid. They can also be in the process of transformation from liquid to solid as they dry up, such as slimy phlegm and coagulating blood, or in transition from solid to liquid (or vice versa), such as muddy melancholy.

Cioran's languid mud of melancholy is a secularised slough of despond with strong theological roots that he acknowledges explicitly in a later section of *A Short History of Decay* devoted to the "Genesis of Melancholy." Melancholy has its beginnings, or "genesis," for him in the fall from paradise as told in the biblical story in the book of *Genesis*. Cioran (2010, 143) suggests that melancholy "harks back to the root of our ruin" in the fall from paradise and he concludes that melancholy is "the poetry of original sin." Although melancholy is not strictly one of the seven deadly or cardinal sins in the Christian tradition, it is linked to the deadly sin of pride or egotistical self-love.[8] Cioran (2010, 110) figures it as the "strangest flower of self-love." It is also linked to the deadly sins of sloth or laziness, and greed or gluttony (making up three out of the seven). It can lead to despair (the one unforgiveable sin), self-hate (anger or wrath is another of the seven deadly sins) and suicide (which is a sin, partly because the perpetrator is not able to ask for forgiveness afterward). Whereas wrath is anger directed against another person, melancholy is anger or hate (and ambivalently pride or love) directed toward oneself, and greed and gluttony directed toward the earth. All these sins flow from the original sin of disobeying God's interdiction against eating the fruit of the tree of knowledge of good and evil. No sin*s* without the singular original sin of knowing sin by consuming the fruit of God's creation. Melancholy is associated with so many other sins that it is a multifaceted mega-sin.

Cioran secularises the fifth circle of Dante's poetic hell in which melancholics, with "black bile billowing" inside them (following the theory of the humors, Pensky 2001, 22–23 and 252, n. 39), are "bitter in the blackened mud" of a swamp (as Dante puts it in keeping with the denigration of wetlands). They have fallen into the slime of a swamp where the souls of the sullen are stuck.[9] The sullen have committed the deadly sin of anger against another and they embody wrath. The melancholic commits the deadly sins of pride and self-anger, and embodies self-love/hate. They are stuck in the same circle as the sullen of the medieval slime of hell for Dante, secularised as the modern hell of slime by Jean-Paul Sartre and others.[10]

Hatred resides in the figurative wetland of what Leonard Cohen, the musical master of modern melancholy, called "the swamps of hate" of racism and sexism. They are linked to the placism of the hatred of swamps (misaquaterrism) that describes wetlands in denigratory terms with their sullen miasmas, poisonous vapours, sucking slimy mud, entangling vegetation, dirty, disease-ridden water, and evil and orally sadistic monsters (imaginary or real, such as alligators and crocodiles) lurking in the depths of their dead black waters waiting to devour greedily the unwary who venture into their watery/earthy domain of slime.[11]

Love, by contrast, lives and thrives in the native quaking zone of living black waters with their monstrous creatures and their fertile and fetid water that invite wading into the slimy swamp when it is safe to do so, a sacred space and place in which to self-baptize and evade the heights and depths of despair, cultivate gratitude for their generosity and faith in life—even in death—and nurture hope (as Henry David Thoreau found, and as we will find in chapter 8 of the present volume). Melancholic ambivalent self-love/hate mediates these two wetlandscapes as Cioran suggests in his figuring of melancholy as languid mud and as morally and aesthetically ambiguous.

The infinite wetlandscape of ambivalent melancholy arises from the obverse of the infinite spatiality of the sublime (see figure 2.1 and Giblett 1996, chapter 2). Without the sublime, melancholy would not be possible. Without the secular theology of the sublime, the fall into the secular demonology of melancholy would not be possible. Without sublimation into the abstract and surmountainous heights of despair, philosophy and theory, the fall into the depths of slime in a secular hell and the rise into the black sun of miasmatic melancholy would not be possible. The sublime is sometimes accompanied with melancholy as Kant said (cited by Giblett 1996, 162). Both sublime and melancholy entail upward displacement, or artificial transcendence. Both accompany each other on upwardly parallel paths, or trajectories as it were (see figure 2.1). Both differ from the infinite temporality, or eternality as in Freud's discussion, of the wetland.

THE GRIM DOMAINS OF GIANT DESPAIR AND MELANCHOLY

Despair and melancholy have been figured in slightly differing wetland tropes with melancholy as the slough of despond and despair as a wetwasteland. In the early modern vision of hell granted to St Teresa of Avila (1957, 233) in a moment of despair "the ground appeared to be covered with a filthy wet mud, which smelt abominably and contained many wicked reptiles." Filthy mud, bad smells, and evil monsters are associated with despair and hell by St Teresa.

Charles Dickens follows suit and does likewise in the nineteenth century with "the grim domains of Giant Despair" in *Martin Chuzzlewit.* Despite the differences in theory between despair and melancholy outlined by Cioran, Dickens conflates them and highlights the difficulty of keeping them apart in practice, though both for him reside in, and are associated with, a wetland. The final leg of the journey of Martin Chuzzlewit to Eden at the confluence of the Ohio and Mississippi Rivers as related in chapter 23 of *Martin Chuzzlewit* is through a scene made up entirely of:

> sky, wood and water, all the livelong day; and heat that blistered everything it touched. On they toiled through the great solitudes, where the trees upon the banks grew thick and close; and floated in the stream; and held up shrivelled

> arms from out [*sic*] the river's depths; and slide down the margin of the land, half growing, half decaying in the miry water. On through the weary day and melancholy night: beneath the burning sun, and in the mist and vapour of the evening: on, until return appeared impossible, and restoration to their home a miserable dream.

This is a journey into a place where the elements are mixed, a place which gives birth to death or to a land where the trees are half-growing and half-dying. This mixture of the elements and life and death gives birth to "the dull depression of the scene." The journey is a secularised epic descent into the underworld on "old Charon's boat, conveying melancholy shades to judgment." Judgment of whom and for what is not specified, but the post-epic journey through the swamp as modern secular hell is part of the test which Martin must complete to achieve manhood on his journey to return home.

From this point the situation can only deteriorate as these are only the approaches to the underworld; this is not Hell itself but only the journey across the River Styx as related in chapter 23 of *Martin Chuzzlewit*:

> as they proceeded further on their track, and came more and more towards their journey's end, the monotonous desolation of the scene increased to that degree, that for any redeeming feature it presented to their eyes, they might have entered, in the body, on the grim domains of Giant Despair. A flat morass, bestrewn with fallen timber; a marsh on which the good growth of the earth seemed to have been wrecked and cast away, that from its decomposing ashes vile and ugly things might rise; where the very trees took the aspect of huge weeds, begotten of the slime from which they sprung, seeking whom they might infect, came forth at night, in misty shapes, and creeping out upon the water, hunted like spectres until day; where even the blessed sun, shining down on festering elements of corruption and disease, became a horror; this was the realm of Hope through which they moved.

Giant Despair is a character straight out of Bunyan's *Pilgrim's Progress* who owns "Doubting-Castle" (Bunyan 2008, 116). Dickens transforms despair into the characteristic of capital "D" Despair of gigantic proportions and transfers the giant's domain from a castle into a dismal swamp and slough of despond.

Although for Dickens Eden is ironically not a paradise but a festering sink in which the elements are improperly mixed and from this improper mixing spring hideous and monstrous deformities and mutants, by a further countervailing irony the wetland chaos, as I suggested in the previous chapter, could be seen to be God's first work, not so much Eden, the pastoral paradise, but the pristine pre-Edenic, pre-pastoral and pre-paradisaical wetland (as depicted in the foreground of Eugene von Guérard's painting, "Mount William and part of the Grampians in West Victoria," figure 5.2). Dickens' Eden, however, is far from that watery chaos. It is both a "hideous swamp" "choked with slime" and a "dreary situation" that produces "dread" and "the apprehension of death." Like the approaches to the swamp-town in which the elements of heat, water, air and earth are mixed, so too in Eden itself they are mixed where "a fetid vapour, hot and sickening as the breath of an oven, rose up from the earth, and hung on everything around."

At Eden as related in chapter 23 of *Martin Chuzzlewit* Martin could not leave his footprints on "the marshy ground" for as soon as his foot sank into it, "a black ooze started forth to blot them out." Martin is also unable to leave much of a mark on the land, or more precisely wetland, as it was "mere forest" whose:

> trees had grown so thick and close that they shouldered one another out of their places, and the weakest, forced into shapes of strange distortion, languished like cripples. The best were stunted, form the pressure and want of room; and high about the stems of all, grew long rank grass, dank weeds, and frowsy underwood: not devisable into their separate kinds, but tangled all together in a heap; a jungle deep and dark, with neither earth nor water at its roots, but putrid matter, formed of the pulpy offal of the two, and of their own corruption.

This wetland is a nightmare for the taxonomizing botanist who would seek to separate out its specimens of flower, fruit and leaf into genus and species. This wetland is not just the excrement of earth and water, but the corrupt, decomposing excrement of earth and water. More generally, wetlands are a taxonomic anomaly in a classificatory order predicated on a hard and fast distinction between land and water, time and space.

MOURNING, MELANCHOLY, AND BENJAMIN

Along Freudian lines, Ilit Ferber (2013, 20) reiterates in a recent discussion of Walter Benjamin's (1977, 2019) work on the German mourning play (*Trauerspiel*) that in melancholy "the lost object continues to exist, now as part of the dejected subject." The lost wetlands destroyed by cities continue to exist, now as part of the dejected subject of the citizen (the denizen of the city) who maintains "a relationship with an absent lost object" (Ferber 2013, 43), just as the subject maintains a relationship with absent lost wetlands by being oblivious to them or by retelling their stories, by remarking their presence in the past on maps and their absence in the present on maps too. The aim thereby is that "the relationship between subject and object is overturned"

(Ferber 2013, 46). Subject and object become what Kristeva (1982, 1–2) calls (the) abject, or what Nietzsche called the body (see Giblett 2008a, 3–5). The subject and object return to the pre-subject and pre-object phase in which both were abject. The abject precedes the subject and the object. It is the third party that made both possible. The dejected subject in no longer dejected, but abjected and is no longer subject, but abject.

This de-subjectification, de-objectification and abjectification contrasts with what Ferber (2013, 59) describes as "the destructive nature of the melancholic's response to loss" in which "the melancholic devours his lost love-object in order to retain it; in demonstrating his endless loyalty, he destroys it. Therein lies the paradox: the only way to retain the object is to destroy it." The melancholic devours his lost love-object of wetlands by using monstrous drainers and dredgers in order to retain them (literally behind retaining walls as in some cities such as Venice) and retrain them and their aberrant water (as in the city of Venice in canals; see Giblett 2016a, chapter 5), but in doing so kills the wetland as a living being, as a habitat for other living beings. He kills the thing he loves, rather than loving the thing he kills as indigenous cultures do (see Giblett 2011, 215). For Ferber (2013, 73 "the melancholic's lost object is either dead (in the case of human loss) or absent (in the case of a more abstract loss)." In the case of the destruction and loss of a wetland, the melancholic's lost object is both dead (a case of nonhuman loss) *and* absent (a case of a concrete loss).

Furthermore, for Ferber (2013, 73), "in both cases the pathology lies in the subject's inability to recognize the loss and the insistence of maintaining the dead object as 'half-alive' within the melancholic consciousness, thus rendering the boundary between life and death indefinite and thus indistinct." In the case of the destruction and loss of a wetland, the psycho-geopathology lies not only within what Ferber outlines, but also within not recognizing the wetland as a place of both life and death in which life and death are mixed, but wherein the boundary between them is definite and thus distinct. The melancholic's half-alive lost object for Ferber (2013, 102) is "what is not yet dead but no longer alive." This object is buried alive in a tomb when, Ferber (2013, 102) goes on to relate, "the melancholic carves out an internal tomb for his lost object, engendering an internal topography in which the living ego and the dead object coexist," just as in Venice the city goes on living above its entombed dead lagoons of once living wombs of wetlands (see Giblett 2016a, chapter 5).

In the concrete cases of the subject's relation to nonhuman living objects, such as animals and wetlands, the living ego and the dead object coexist by the living ego continuing to live by killing the object (see Giblett 2008a, 119–120; Giblett and Tolonen 2012, 41 and 45). They are entombed as dead object, but they were once life-giving wombs. Death, as Ferber (2013, 104) concludes, "does not mark the end of life . . . : the two states exist concomitantly," as they do in the wetland's ecology. Moreover, for Ferber (2013, 115) "the dead are never completely dead, and the past can never be hermetically closed." The lost wetlands are never completely dead, and the past is still an open book, as it is in Venice (see Giblett 2016a, chapter 5). Venice is the wetland city par excellence not only in terms of its past and its history as a city founded in a swamp, but also in the present and its current culture as a rich source of metaphor and of psychological travail in which these metaphors are symptomatic of its mournful and melancholic psychogeopathology.

MOURNING, MELANCHOLY, AND BRITTAIN

Melancholy and mourning in association with the city of Venice are played out extensively in Vera Brittain's autobiographical study, *Testament of Youth*. As she "glides smoothly" in a gondola "over the rippling grey silk of the Grand Canal" on her post–First World tour of Europe, the view of Venice for Brittain (1978, 479) is tinged with melancholy and mourning as she associates the Venetian waters with the death of her fiancé in France and brother in Italy during the war. The Grand Canal is associated not only with melancholy and mourning, but also with magic as Brittain "with melancholy possessiveness [. . .] looked upon those enchanted waters . . . those fairy lagoons, incredible as a gorgeous mirage in the muffled silence." These are the benign waters of the patriarchal mother earth. She imagines that her brother "had died saving this beauty from the fate of Ypres," in other words, saving the beauty of the city of Venice from the horror of trench warfare, especially its mud.

Brittain read about the mud of trench warfare in the letters her brother and fiancé wrote to her and she had immediate experience of it as a nurse in France and in the aftermath of the death of her fiancé. In one of his letters from the trenches Brittain's fiancé related that they were "very wet and muddy . . . The whole of one's world, at least one's visible and palpable world, is mud in various stages of solidity or stickiness" (Brittain 1978, 206). Mud mixes the elements earth and water, whereas cities and roads separate them as a rule, or should do. Brittain (1978, 337) experienced the mud first-hand when she arrives on the Western Front and found that "the roads were liquid with such mud as only wartime France could produce after a few days of rain." The solidity of earth was liquefied into mud.

The mud of trench warfare, especially its smell, in associated for Brittain with death when she goes through the "kit" of her fiancé following his death. She wrote to her brother that everything was "simply caked with mud" and she felt "overwhelmed by the horror of war" as "the smell of those clothes was the smell of graveyards and the Dead. The mud of France which covered them was not ordinary mud; it had not the usual clean pure smell of earth, but it was as though it were saturated with dead bodies—dead that had been dead a long, long time" (Brittain 1978, 225). The mud of trench warfare had the smell of death in which earth, water, and dead bodies (and shit [though she does not mention this]) were mixed. It did have not the usual smell of mud in which earth and water were mixed. Brittain distinguishes the clean, pure smell of earth from the horrific, dirty smell of the mud of trench warfare.

Like most writers about the World War I (see Giblett 2009, chapter 4), Brittain (1978, 355) associates the mud of trench warfare with swamps and marshes, such as when she describes how "the terrific gales and whipping rains of late autumn . . . turned the shell-gashed flats of Flanders into an ocean of marshy mud that made death by drowning almost as difficult to avoid as death from gun-fire." Yet the wetlandscape of trench warfare was an artificial marsh made by modern industrial warfare, not a native marsh made by hydrogeological processes and ancestral hands. It was a feral quaking zone, not a native quaking zone.

The wetlandscape of trench warfare is also melancholic for Brittain (1978, 356), such as when she describes how "the Flanders offensive was subsiding dismally into the mud" and refers to "melancholy Flanders" on the next page. Brittain experiences mourning and melancholia during and after the World War I, yet she also experiences these affects in Venice after the war in combination with the magic of the maternal waters of the wetlands—perhaps hardly surprising given her strong feminist beliefs, though she is unable to distinguish the canalized waters of the patriarchal city from the matrifocal waters of the wetlands that preceded the city and which the city destroyed to be built. Although she distinguishes the dirty impure smell of trench warfare from the clean pure smell of earth, she does not regard the former as the product of patriarchal and industrial trench warfare and the latter as the progeny of maternal marshes, nor does she distinguish the artificial "manmade" marshes of modern industrial warfare from the native marshes made by hydrogeological processes and ancestral hands.

Instead of seeing the object (the breast, the wetland) as lost, the melancholic ego sees itself as lost in a massive act of narcissistic disavowal and egotism. As a result, the ego desires itself, or in the terms of Deleuze and Guattari (1977, 26) "it is the subject [. . .] that is missing in desire, or desire that lacks a fixed subject; there is no fixed subject unless there is repression." Repression is to subjectivity as drainage is to wetlands; repression (and drainage) fixes the flows of embodied subjects (and wetlands). Repression is constitutive precisely of *melancholic* subjectivity; subjectivity is melancholic and mournful. The subject desires itself as a product of a melancholic loss of the loved object of the mother and mother earth, of the mother's breasts and mother earth's breasts, the living waters of wetlands.

As wetlands are increasingly lost from the world and are lost as an object of love which nourishes life, both mourning and melancholia are experienced and exercised in relation to them. The world should be in mourning for the loss of its wetlands which gave it life and nourished it, but instead of being in mourning and regarding the world as losing its wetlands becoming empty or "bereft" of them, in Gerard Manley Hopkins (1953, 51) apt word, it experiences this loss as a melancholic loss of its own ego, its own selfhood, and sense of identity. Selfhood and identity are constituted in relation to something outside itself against which it stands in contradistinction. When that object is lost, as when wetlands are lost, that loss is experienced in melancholia as a loss of the subject. The world is becoming empty of wetlands; the world is losing its wetlands. In patriarchy (the rule of the fathers) in the past and filiarchy (the rule of the sons) in the present wetlands are the bad breast of the world in which, in Melanie Klein's terms, the infantile sons try to hide their excrements, but in these psycho-ecological terms they are the womb and breast of the world, the organs of receptivity and bounty as Klein puts it[12]—they are living waters and living waters matter.

NOTES

1. For discussion of the Great Dismal Swamp on the border between Virginia and North Carolina and its writers, beginning with William Byrd in the eighteenth century, see Giblett (1996).
2. For discussion of the greed and gluttony of mining as the psychogeopathology of anal sadism directed toward the great mother earth, see Giblett (2011, chapter 9; 2019, chapter 5).
3. Renee Aron Lertzman (2013, 124–126) in her discussion of mourning and melancholia in Freud's essay "On Transience" (and not in his essay on the topic) considers "the lost or damaged object," such as "a body of water," as outside and separate from the subject, and not as (part of) the subject. Lertzman (2015, 117) also invokes 'Freudian melancholia' in her book-length study of environmental melancholia, but does not discuss Freud's essay on mourning and melancholia which seems like a lost opportunity to engage with their psychoanalytic dimensions and to relate them to environmental issues and concerns.
4. The association between blue and melancholy is discussed further in chapter 7 of the present volume.
5. For the landscapes of world warfare, see Giblett (2009, chapters 4 and 5).
6. For further discussion of Venice as a wetland city, see Giblett (2016a, chapter 5).
7. Despite bidding "farewell to philosophy" when "it became impossible to discover in Kant and in all the philosophers any human weakness, any authentic accent of melancholy," Cioran (2010 49, 54–55) still entertained enlightenment philosophical

metaphors, such as when he claims we "habitually we sink into a nocturnal mud, into a darkness quite as mediocre as the light [. . .] Life [. . .] is a monotonous loss of light, an insipid dissolution in the darkness [. . .]." There would be no enlightenment/Enlightenment without a corresponding and concomitant process and movement of "endarkenment," in Cioran's case, of the muddy and living black waters in the heart of darkness. Cioran still turned toward philosophy in turning away from it. He was Janus-faced, part of his fascination.

8. Pensky (2001, 31) suggests that "as a cardinal sin, melancholia proceeds as a direct consequence of the Fall." The former assumption is mistaken I suggest, while the latter diagnosis concurs with Cioran.

9. For further discussion of Dante and slime, see Giblett (1996, 27–29).

10. For further discussion of Sartre, slime and the swamp as secularised hell, see Giblett (1996, 39–47, 138–145).

11. For further discussion of alligators and crocodiles as uncanny creatures of the wetland, see Giblett (2019, chapter 3).

12. For further discussion of Klein and mining as the psychogeopathology of anal sadism, see Giblett (2011, chapter 9; 2019, chapter 5).

Part II

WETLANDS, ART, AND CULTURE

Chapter 3

Wetland Cultures of the English Fens

Politics, Painting, Poetry, Prose, and Art History

Recent work on the English Fens coming out of art history, landscape aesthetics, and landscape architecture demonstrates the still contentious and problematic positioning of the Fens in English culture, the need for concerted efforts to conserve them and the value of recent work to restore them (Hillsdon 2020). Vittoria Di Palma's (2014) lavishly illustrated and eponymously titled art history of wasteland devotes a chapter to the English Fens that critiques their enclosure and draining, appreciates their unpicturesque representation in landscape painting and traces the development of artificial wetlands in landscape architecture. Firmly located within the modalities of visual representation and the disciplines and discourses of art history and landscape aesthetics and architecture, this study neglects or overlooks the miasma, malaria and melancholia of the Fens as diagnosed and discussed in medicine, philosophy, and psychology. These topics and disciplines are included in a more transdisciplinary approach to the Fens engaging with cultural, historical, and literary studies in the environmental humanities in *Postmodern Wetlands* (Giblett 1996).

Other disciplines, such as archaeology, and other arts, such as church architecture and construction (including its legendary carpentry), are extensively and thoroughly considered by Francis Pryor (2019) in his personal and professional memoir of forty years of study of the Fens' built structures and agricultural practices in ancient, medieval and modern times. Recently the Fens have become the focus and site of "re-wetting" and recreating "paludiculture" with "a 100-year plan to re-wet 3,700 hectares of peat across Cambridgeshire," plant more than 200,000 plugs or tube-stocks of commercially viable native grasses, rushes, and mosses, halt the loss from the Fens of "4.5 million cubic metres of peat a year to soil erosion" and conserve water in the land for farming and wildlife. It is "one of the largest restoration projects of its kind in Europe" (Hillsdon 2020).

Building on *Postmodern Wetlands*, drawing on Di Palma's work, and interweaving Pryor's and other recent work on the Fens in painting by Fenland artist Nick Tearle (n.d.) and verse by Fenland poet Becky Owen-Fisher (2017), the present chapter takes a political approach to the Fens and the enclosure of the commons into private property critiqued by John Clare, Karl Marx, E. P. Thompson and Raymond Williams. It also takes a cultural, literary and historical approach by considering the depiction of the Fens in painting and the literature of the Fens, beginning with Fenland native Clare's poetry and prose in the nineteenth century for the Fens and against enclosure. It then proceeds to a consideration of the Fens in later prose, beginning with Dorothy Sayers' classic detective novel *The Nine Tailors* (1934), Graham Swift's wetlandmark *Waterland* (first published in 1983) and his 25th anniversary introduction of 2008, Phillip Pullman's *Northern Lights* (1995), the first part of his trilogy, *His Dark Materials*, Robert Macfarlane's (2008) "new nature writing" of "Ghost Species," culminating recently with Becky Owen-Fisher's poetry (2017) and Stella Tillyard's historical romance *The Great Level* (2018).

THE FENS IN ART HISTORY

In her chapter on the Fens in *Wasteland*, Di Palma (2014, 95) argues that "the swamp is a landscape that is dangerous not only to the body, but also to the soul." The Fens were considered dangerous to the body because of their bad airs (as Di Palma discusses, but she does not go on to point out the reason for this view). The Fens were considered dangerous according to the miasmatic theory of disease that contended with the germ theory of disease in medical theory and practice up to the late nineteenth century. The swamp is dangerous to the body because ague (or malaria—literally "bad air") was thought to be contracted by breathing the miasma or bad air rising from swamps, marshes, and other stagnant pools of water, including in cities, especially slums. Nor does Di Palma discuss the disease of ague (or malaria) that was commented upon by many writers who bemoaned the unhealthiness of the Fens and other wetlands around the world in accordance with the miasmatic theory of disease, instances of which could be enumerated almost ad infinitum (see Giblett 1996, chapter 5).

The Fens, and other wetlands, were not only bad for the body, but also bad for the soul, or for mental health, in the European philosophy and psychology of melancholia. Di Palma (2014, 19–22, 95) focuses her account of this aspect of wetlands on John Bunyan's *Pilgrim's Progress* in which as the scum and filth of sin is figured as "the slough of despond" that would drag Christian down and engulf him for all eternity. The slough of despond is akin to the marsh of melancholy (as we saw in the previous chapter). Like malaria, melancholy was associated with the miasmatic theory of illness. The miasmatic theory of melancholy arose with Robert Burton and John Bunyan in the seventeenth century, and went back through the Elizabethans of the sixteenth century, to the Greek philosophy of the four elements and humors (or psychosomatic states) that miasmatic gases caused melancholy (see Giblett 1996, chapter 7). It went hand in hand with the miasmatic theory of malaria.

John Bunyan's *Pilgrim's Progress* was first published in 1678. Over half a century earlier in 1621, Robert Burton published *The Anatomy of Melancholy*. It was republished four times within the succeeding seventeen years. Bunyan in *Pilgrim's Progress* was not only singing from the song-sheet of *Psalm* 23: 4 in depicting the terrors of walking in "the valley of the shadow of the death" (as Di Palma 2014, 20 notes and as we also saw in the previous chapter), but also sermonizing in accordance with Burton's anatomizing of melancholy (as Di Palma does not note) in warning about the horrors of "the slough of despond." Miasma, malaria, and melancholia are missing from *Wasteland*. They were part of the moralization of the Fens and other wetlands, and contributing factors in their drainage.

THE FENS IN POLITICS

Coming out of art history and lavishly illustrated with paintings, plans, maps, and sketches, Di Palma's (2014) landscape history of wasteland devotes an entire chapter to "Swamp," or more precisely, to the Fens of England. It documents some appalling lowlights in the history of the English Fens, such as the class robbery perpetrated by their enclosure that went hand in hand with the ecocide, or more precisely the aquaterracide, of their draining (Di Palma 2014, 25, 35, 45, 52, 58, 66, 99–112). In the middle of this conjunction and discussion, Di Palma invokes Karl Marx's scathing critique in chapters 27–31 of the first volume of *Capital* of the enclosure of "the common land" (dryland and wetlands) into private property (Di Palma 2014, 55–56; Marx 1976, 877–926). Enclosure in short was what the late, great E. P. Thompson (1968, 237) called "a plain enough case of class robbery."

Marx argues that the enclosure of the commons and "the clearing of the estates" (a euphemism for the dispossession and driving-off of the commoners, or local people) was the foundation for industrial capitalism and modern colonialism. In the final chapter (33) of *Capital I* entitled "The Modern Theory of Colonization" Marx concludes that "the expropriation of the mass of the people from the soil forms the basis of the capitalist mode of production" (Marx 1976, 934). It is a cruel irony of history that some Scottish settlers of western Victoria in Australia who had been subjected to "the clearing of the estates" in Scotland later "enacted their own brutal clearances on Indigenous [Aboriginal] peoples" (Wilkie 2020, 46). Kiddle (1961, 14, 517 n1) calculates that "at least two-thirds of the pioneer settlers of the Western District [of Victoria] were Scottish. Nearly all were lowland farmers."

A lack of sympathy or empathy from them toward Aboriginal people also did not extend to their acknowledging a similar clan culture in structure, even though some Scottish settlers of western Victoria recognised the similarities between the two societies (Wilkie 2020, 21–22). This did not lead to the settlers expressing any fellow-feeling or exercising solidarity toward Aboriginal people. It was more a case of repetition and displacement, of repeating dispossession and clan antagonism against Aboriginal clans. The settlers were doing to others what had been done to them elsewhere and previously (rather than obeying Jesus' second great commandment of doing to others what you would have them do to you).

As an aside, Marx's communist critique of the enclosure of the commons is strangely neglected or overlooked by Guy Standing (2019) in his recent, timely, otherwise laudable and politically progressive manifesto, *The Plunder of the Commons.* Indeed, Standing (2019, 43) lumps together "communism, socialism, and liberalism" and asserts that they "all viewed the commons as dispensable while economic growth was sacrosanct. Progress allowed little respect for tradition and common rights." This was certainly *not* the case with communism according to Marx who viewed the enclosure of the commons as dispossession, not as dispensable, and as a violation of tradition and common rights (as we have seen); it is the case with the "state capitalist" countries, or centralised, one party-controlled economies of so-called communist nation-states. Standing does not specify that he is referring to the latter. By contrast, Bregman (2020, 308) in his heart-warming and hopeful history of humankind works with the former by defining "communism" as "the bedrock on which everything else—markets, states, bureaucracy—is built."[1]

The enclosure of the commons into private property is an aspect of Marx's (and Thompson's) work on capital and ideology also strangely neglected or overlooked by Thomas Piketty (2020) in his recent, sweeping, otherwise laudable and politically progressive study, *Capital and Ideology.* Indeed, Piketty (2020, 53) argues that "the two ruling classes [in ternary societies]—clergy and nobility—were of course propertied classes. They generally owned the majority (and sometimes nearly all) of the cultivable land, which is the basis of economic and political power in all rural societies," as political economists argued. This majority ownership is probably true of France, but not of England, even though Piketty (2020, 175) later acknowledges "the famous Enclosure Acts" passed by Parliament in

England, "most notably in 1773 and 1801," whose "purpose was to put hedges around fields [! *sic*] and put an end to [the] right of poor peasants to use communal land for crops and pasturage." Their purpose was not only to landscape fields into aesthetically pleasing and picturesque pastoral prospects, but also to enclose politically and economically common land owned in common and shared by all, including the Fens and other wetlands (often by drainage), into private property and so to create "the great landed estates" (Piketty 2020, 451) from which "the peasants" were "cleared." The aesthetic, political, and economic were mutually reinforcing, as Raymond Williams (1973, 124–125) argued, to position and organize the land for profitable agricultural production and pleasurable aesthetic consumption. The intended consequence was to dispossess and drive off the traditional owners and users (reduced to mere peasants in Piketty's view which does not have the same pejorative overtones in French culture as it does in English culture).

In the British enclosure movement "from the second quarter of the eighteenth century to the first quarter of the nineteenth century," as Williams (1973, 96) puts it, "by nearly four thousand Acts, more than six million acres of land were appropriated" (though much of this land was specifically *wetland*, a crucial distinction[2]). Williams (1973, 96) goes on to point out that "the process had been going on since at least the thirteenth century, and had reached a first peak in the fifteenth and sixteenth centuries." The enclosure movement in wetlands was invariably associated with drainage to render the land profitable for monoculture with, as Christopher Hill (cited by Giblett 1996, 47–48) maintains, "fen drainage alone extending the country's arable area by 10%." In fact, the new outburst of drainage projects of the 1760s was contemporaneous with and supportive of the beginning of the agricultural revolution.

Marx's communist critique of the enclosure of common land into private property and of political economy by indicating that this was the basis for the capitalist mode of production is a salutary instance of how environmental conservation is a working-class and indigenous concern of local peoples preserving livelihoods in local places on local land in bioregional home-habitats by using earthly resources, not just a middle-class affectation of "setting aside" special places in sanctuaries for their leisure, recreation, and entertainment (for a critique of "sanctuarism," see Giblett 2011). Marx's work provides a platform for the "socialist ecology" built upon later by Williams as a pioneer of the environmental humanities and the patron saint of eco-cultural studies in his trenchant critique of the enclosure of the commons, and in his work more generally on nature, the country and the city, landscape, livelihood, and resources for a journey of hope (for further discussion, see Giblett 2020d, chapter 10).

The expropriation of the soil from indigenous and local owners forms the basis of the capitalist mode of production and colonization in the English colonies where indigenous peoples' lands were taken by squatters (later legitimized after the fact) and granted freely to settlers and where they were the beneficiaries of this class robbery. At home and abroad, endo- and exo-colonisation involved enclosure of the commons and "clearing of the estates," in the case of Australia, "the biggest estate on earth" (Gammage 2011). What Marx (1976, 940) called "the shameless squandering of uncultivated [*sic*] colonial land on aristocrats and capitalists by the English government" occurred not only in Australia (as Marx goes on to indicate), but also in Canada. Colonial land was not, of course, uncultivated, but had been cultivated by indigenous peoples for tens of thousands of years in the case of Australia and Australian Aboriginal peoples (Gammage 2011; Pascoe 2014; Mayes 2018). Class expropriation of local and indigenous people and their lands (wet and dry), and class warfare against wetlands, has gone hand in hand with the draining, filling, and destruction of wetlands ever since in England and in their colonial diaspora in Australia, Canada, and the United States.

THE FENS IN PAINTING

Di Palma's *Wasteland* not only critiques the enclosure and draining of the Fens, but also presents some curious artefacts in the history of wetlandscape architecture, such as the design plans for the first artificial wetlands of "duck decoys" to attract and "harvest" water fowl by shooting them (Di Palma 2014, 115–119; see also Pryor 2019, 168–169[3]). *Wasteland* not only presents these appalling lowlights, but also showcases some appealing highlights in art history and landscape painting. English wetlands have not figured prominently in English landscape painting as "the marshy Fens," were, as Di Palma (2014, 123) puts it, "the epitome of the unpleasing, unpicturesque landscape," especially for such theorists and practitioners of the picturesque as William Gilpin. Wetlands were also neither sublime nor beautiful for such poets as William Wordsworth (see Giblett 2011, chapters 3 and 4).

For Di Palma (2014, 125–127) the epitome of the unpicturesque wetlandscape in paintings of the Fens are two gloomy depictions by John Sell Cotman (1782-1842). One is entitled "Drainage Mills in the Fens, Croyland, Lincolnshire" (Cotman, *ca*1835). The other is entitled "St. Benet's Abbey" and is attributed to Cotman (*ca*1810). Both paintings are quite atypical of his entire body of work. In the catalogue for the exhibition held for the bicentenary of Cotman's birth in 1982, none out of over 130 works listed and discussed are of the Fens. Admittedly, "the catalogue entries do not attempt to be exhaustive" (Rajnai 1982, 7). Indeed, none of the considerable holdings of Cotman's work in the Yale Center for British Art in New Haven, Connecticut, including the two paintings discussed by Di Palma, are included in this catalogue. The two paintings of the Fens in the Yale Center are also quite atypical of this collection and of the 275 holdings in the Cotman collection in the Yale Center for British Art (n.d.). Cotman is hardly the painter of the Fens.

A perhaps more likely contender for this accolade is Cotman's contemporary Peter de Wint (1784–1849) whose work was appreciated by John Clare (as we will see below) and recently by Pryor (2019, 391–392). De Wint painted two similar paintings both depicting the low-lying wet landscape of the Fens as the background to a scene of heroic agricultural labor with the archetypal, if not clichéd, presence of a haywain mythologised by Constable as a symbol of taming the land and deriving benefit from it. One of these paintings is entitled "Lincolnshire Cornfield near Horncastle" and is held in the Usher Gallery in Lincoln (De Wint, 1815b); the other is entitled "A Cornfield" and is held in the Victoria and Albert Museum in London (De Wint 1815a).

De Wint painted other "low-lying landscapes in the Dutch manner" (Scrase 1979, 11), but none of these paintings are of the low-lying landscape of the Fens, a low-lying wetlandscape. They were primarily of rivers as he was "partial to river scenery," according to his wife Harriett (cited by Scrase 1979, 11). In other words, he was not partial to Fenland scenery, or perhaps more precisely, he was not partial to Fenland obscenery in the sense of what is hidden and secret. Besides river scenery, he was also partial to wifely scenery as he painted a stunning, erotically charged portrait of his wife with diaphanously veiled breasts, probably for his own private viewing pleasure and held in the Usher Gallery in Lincoln (De Wint n.d.). He did not paint landscapes with the same amount of passion, certainly not the Fens as they are largely absent from De Wint's entire oeuvre. They remain hidden and secret, unlike his wife's upper body. None of the one hundred drawings and watercolors listed and discussed in the catalogue for the exhibition held in the Fitzwilliam Museum in Cambridge in 1979 are of the Fens (Scrase 1979). De Wint did not depict premodern and mundane low-lying wetlandscapes, or when he did (such as in the two paintings mentioned) it was only as a background to agriculture.

THE FENS IN POETRY

The poet John Clare (1793–1864) was a contemporary of De Wint. Clare is for Pryor (2019, 77) "the greatest poet of the English landscape." He is also the greatest poet *for* the Fens and *against* enclosure. In Clare's poem,

"The Fallen Elm," ruin was the guide of enclosure (cited by Pryor 2019, 79; Standing 2019, 14). Clare goes on to describe how "The common heath became the spoilers prey" (Williams and Williams 1986, 85). In "The Mores" and "The Lament of Shady Well," Clare goes on further to describe how "vile enclosure" "came and trampled on the grave/Of labours rights and left the poor a slave," including Clare himself (Williams and Williams 1986, 91 and 98; cited by Standing 2019, 43). In "Remembrances," enclosure, like a Napoleon, "let not a thing remain" (Williams and Williams 1986, 150). Enclosure was war against wetlands, the commons and their peoples, the commoners. For Clare, "where enclosure has its birth/Its spreads a mildew o'er her mirth" (cited by Williams 1973, 136). Where the elm once stood are now "workhouse prisons raised upon the site" (cited by Pryor 2019, 79). Enclosure, "thou'rt a curse upon the land" (cited by Williams 1973, 136). Clare is specifically the greatest English poet against enclosure and for the Fens who wrote two long poems, "The Fens" and "To the Snipe" addressed to "lover of swamps" (Williams and Williams 1986, 109–112, 150–153). In their commentary on Clare's poetry and prose, Williams and his daughter describe how "the fens, as the poem of that name illustrates, are swarming with life[4]" (Williams and Williams 1986, 209). Clare is the patron saint of the Fens (Giblett 2020d, 34 and 140).

Clare praised De Wint for not indulging in "modern fancy landscapes" (cited by Barrell 1972, 146). In a sonnet addressed to De Wint, Clare was more specific about what he meant by "modern fancy landscapes" when he waxed lyrical that:

> [. . .] in thy landscapes I can well descry [. . .]
> No painted freaks no wild romantic sky
> No rocks or mountains as the rich sublime (cited by Scrase 1979, 52)

Rather, what Clare found in De Wint's landscapes was "the sunny truth/Of nature" (cited by Scrase 1979, 52). He did not find the Fens depicted as the gloomy truth of nature (as he might have found in Cotman's paintings), nor did Clare find in De Wint's landscapes reeds and Fens as the fertile slime, nor "swamps of wild rush-beds and sloughs" squashy traces' that he evoked in one of several short poems entitled "Song" (cited by Barrell 1972, 148). Clare may have found the wild romantic sky blue in the painted frock and the rolling hills—the pleasing prospect, as it were—of the breasts of De Wint's wife in his portrait of her, but he does not say. De Wint was not the painter of the Fens, however much Clare admired his work. Perhaps Nick Tearle, the self-proclaimed "Fenland artist," is the painter of the Fens (as we will see below).

THE FENS IN PROSE

The English Fens have not figured prominently in English nature writing with the notable nineteenth-century exception of John Clare's poetry and the recent exceptions of Robert Macfarlane's nonfiction essay, "Ghost Species," (2008) and Becky Owen-Fisher's poetry (Owen-Fisher and Tearle 2017). The Fens have figured more prominently in English prose fiction with Dorothy Sayers" *The Nine Tailors* (1934), Graham Swift's *Waterland* (2010b) and Stella Tillyard's *The Great Level* (2018) with all these novels being set in whole or part in the Fens.

Dorothy Sayers' "great detective thriller" *The Nine Tailors* for Pryor (2019, 50) could "only have been written by someone who knew the fen landscape and people intimately." In fact, Sayers "spent much of her childhood and early adulthood" in a Fenland village. It is hardly surprising then that Pryor (2019, 50) concludes that "her affection for the place comes across vividly." Early in *The Nine Tailors* Sayers (1934, 4) also paints a vivid, critical picture of Fenland drainage when she describes how "the drain ran straight as a rule could make it, black and sullen, with a steep bank shelving down to its slow, unforgiving waters." Similarly, Robert Macfarlane (2008, 109) recently remarked of the Fens how "the landscape becomes rectilinear: ruler-straight road and field edges." The straightness of the roads and dykes is the stock-in-trade description of the drained Fens in *The Nine Tailors* and fifty years later in Graham Swift's *Waterland* first published in 1983.

In depicting a scene reminiscent of Cotman, Sayers (1934, 5) goes on to describe how, "at the end of a solitary mile, the gaunt shape of a windmill loomed up upon the farther bank of the drain, but no bridge led to it, and no light showed." The windmill is imbued with menacing qualities, a monstrous emanation of a dark place. The gaunt windmill contrasts with the awe-inspiring church that Sayers (1934, 34–35) describes in much more intimate and loving detail than she does the Fens. As the daughter of a rector of the church in a Fenland village (Pryor 2019, 50), she was well-trained in architectural description, but also in theological exposition (as we will see later).

Of course, the Fens are not necessarily a pleasant place. One character "with a faint London twang" describes them as "not a nice part of the country" (65) and much later says "it's a hell of a country" (324). Perhaps he is displaying the city-dwellers" prejudice against wetland country, city heaven versus wetland hell, and waste land. Sayers" (1934, 5 and 38) narrator evokes the Fens vividly in winter as "this frozen desolation," "the flat, white wastes of the Fens, [with] the spear-straight, steel-dark dykes."

By contrast, Nick Tearle's paintings, "Winter at the Mille, West Deeping" and "Inkerton Fen," convey the flat whiteness of the Fens in winter, while his paintings, "Ditch Near Etton," "King Street Drain" and "Narrow Drain," show the spear-straight, blue-light ditches and drains reflecting the sky (Tearle n.d.). The Fens for Tearle aren't wastes, though they are dark as in "Rain on Morton Fen" and light as in "Autumnal Sunset on Straight Drove," "Sunset over Six Score Road," "Fenland Morning," "Sunrise Near West Deeping," "Fenland Sunrise" and "Sunset in the Fens," and in-between as in "Fenland Twilight" (Tearle n.d.). The Fens come out into the open and no longer remain hidden and secret in the paintings of Nick Tearle, the self-proclaimed "Fenland artist," some of which accompany Becky Owen-Fisher's poetry (Owen-Fisher and Tearle 2017). Tearle conveys the changing light of different seasons and weather in the "low flatlands" and "flattened landscape" of the Fens that "we call home" (Owen-Fisher and Tearle 2017, 3, 5).

Along similar lines to Tearle's paintings of the Fens in spring, such as "April on Langtoft Fen" (Tearle n.d.), in *The Nine Tailors* Sayers' (1934, 69) narrator later evokes the Fens vividly in spring when, "in its own limited, austere fashion, the Fen acknowledged the return of the sun. The floods withdrew from the pastures; the wheat lifted its pale green spears more sturdily from the black soil." The Fens are fertile and well-watered in a wetting and drying seasonal cycle, the very qualities that make them so agriculturally productive. The Fens are no longer a dark, desolate, white wasteland, but a light, fecund, green wetland when "the green fen spread out mile upon mile" (Sayers 1934, 74). Summer can be a different story, though, when "that summer the water lay on the land [. . .] and the sodden ricks took fire and stank horribly" (369). With the eventual return of winter even the usually unflappable Wimsey remarked that "they seem to keep a special brand of disgusting weather in these parts" (371).

Sayers has her most extensive discussion of the drainage of the Fens come from the mouth of her fictional detective, Lord Peter Wimsey, who describes some sections as "rather more effectively drained" (63). Her most trenchant critique of drainage also comes from him as he describes some sections as "bad engineering" (64). Sayers (1934, 197) saves her most prescient critique of drainage of the Fens until midway through *The Nine Tailors* when Wimsey remarks:

> If this country had been drained intelligently and all of piece [. . .] by running all the canals into the rivers instead of the rivers into the canals, so as to get a good scour of water, [the village] might still be a port and the landscape would look rather less like a crazy quilt. But what with seven hundred years of greed and graft and laziness, and perpetual quarrelling between one parish and the next, and the mistaken impression that what suits Holland must suit the Fens, the thing's a mess.

Mess indeed when land and water are out of place and in the wrong place. This is the kind of critique of drainage that the reader might expect to find more in a politically green environmental history of the Fens than in detective fiction.

The messiness of the Fens certainly comes to the fore in Sayers' (1934, 372) compelling description of a flood in the Fens, an invariable consequence of canalising rivers, rather than vice versa:

> Dark and menacing, the swollen flood-water raced through the sluices, eddying and turning and carrying the brown reeds and broken willow-stems and here and there fragments of timber filched from the drowned lands of the Upper Fen. And even while Wimsey watched, there came a change. Angry little waves and gurges ruffled the strong flow of the river, with an appearance as of repressed tumult and conflict.

The archaic "gurges," related to gorge as both ravine and throat, neatly combines a hydrological reference to drowning in a swirling eddy with an anatomical reference to swallowing in a monstrous act of oral sadism. Platt's (2019) account of "the drowned world" across England and Wales in "the great flood" of 2013–2014 follows in the muddy footsteps of Sayers' vivid description.

The flood is both a return of the repressed of the canalized water and a return to the repressed when the Fens were wetland. Sayers (1934, 373) goes on to describe:

> Dyke and drain were everywhere abrim and here and there the water stood in soaked fields as though they needed but little more to sink back into their ancient desolation of mere and fen.

Or should that be, "merge back into their indigenous elation of mere and fen"? Just when Sayers seems to be going to make her most telling historical and ecological critique of drainage of the Fens, she retreats to a triumphalist chronology of ancient versus modern, waste land versus drained land.

Sayers (1934, 374) also retreats into a Christian theology of the wetland when she repeats the tried and true trope of the melancholy marshes: "the air was so heavy with water" the sound of church bells "drifted through the streaming rain with an aching and intolerable melancholy, like the noise of the bells of a drowned city pulsing up through the overwhelming sea." On the second page of *The Nine Tailors* "the sound of a church clock" prompts Wimsey to expostulate, "thank God! Where there is a church, there is civilization" (Sayers 1934, 4). Where there is no church, as in the pre-Christian Fens, there is barbarism and paganism (or indigeneity and sacrality, I would prefer to say). Where there is a muted church, there is the melancholy marsh and the slough of despond. Where there is the muffled sound of church bells, there is a drowned civilization reminiscent of Atlantis and Richard Jefferies" *After London* and prescient of J. G. Ballard's *The Drowned World* and George Turner"s *The Sea and Summer.*

By contrast with Sayers, in Becky Owen-Fisher's poetry the sound of church bells is: Hitting the dull thud of fogged mystery,

Calling, softly, straining against the wet darkness.
They tell me to follow,
Feet stumbling after their peal,
Hands groping dank empty air [. . . .]
And the singing bells bring me through the bog [. . .]
To the welcome booming peals. (Owen-Fisher and Tearle 2017, 12–13)

For Owen-Fisher, the sound of the bells in the Fens is a friendly guide in a foggy mystery, not a sign of civilization in a hell of a country, nor a frozen waste land nor in a melancholy marsh.

In *The Nine Tailors* the church is not only a sign of civilization, but also a refuge in time of flood as it stood on "a little mound rising some ten or twelve feet above the level of the village, an elevation sufficient to save it in ancient times from inundation during the winter months" (Sayers 1934, 58). In time of flood in modern times, the church is an ark for the salvation of the Christian flock (animals had to fend for themselves) with the rector as latter day Noah. "The drowned Fen" meant that "the whole world was lost now in one vast sheet of water" (except for the ark/church) in a reprise of the biblical flood of gargantuan and global proportions. You can take the rector's daughter out of her father's church and rectory, send her to Oxford, commission her to translate Dante's *Divine Comedy* for Penguin classics, employ her in an advertising agency and get her to write Christian letters to a post-Christian world, but you can't take Christianity out of the detective writer. Like her fellow erstwhile Inklings in C. S. Lewis and J. R. R. Tolkien, Sayers' fiction was a vehicle for evangelizing Christianity.

The result, or down-flow of the flood, is that "the Fen had reclaimed its own" (Sayers 1934, 390 and 391). Or more precisely, the water of the drained and reclaimed Fens had re-reclaimed the Fens. The result is the return of, and to, the repressed wetland with all its glorious, messy, moralized and odiferous features of "the dismal strand of ooze and weed" (Sayers 1934, 395).

Fifty years later Graham Swift's *Waterland* (2010b), "his great novel of Fenland life" for Edward Platt (2019, 133), was published in 1983 to critical acclaim and commercial success culminating in the film of the book starring Jeremy Irons. Crick, the first-person narrator, acknowledges like Sayers, the smell of the Fens, "which is smelt over and over again in the Fens. A cool, slimy but strangely poignant and nostalgic smell" (11). Sliminess is a dominant feature of the Fens, as is flatness. Crick, like Sayers' narrator, offhandedly relates now "the land in that part of the world is flat. Flat, with an unrelieved and monotonous flatness," its "uniform levelness broken only by the furrowed and dead-straight lines of ditches and drains" (10). The Fenlanders' answer to the question, "why are the Fens flat? is typically "so God has a clear view" (21). It "a landscape which, of all landscapes most approximates to nothing" (21). In this "nothing-landscape [. . .] Fenlanders do not try to hide—since we know God is watching" (58).

Similarly, Owen-Fisher emphasizes the flatness of the Fens. In a poem entitled, "The Flatland," Owen-Fisher (2017, 17) describes "saturated ditches" as "Brimming with sleek unreal creatures,/A waterworld of history, bubbling under." This history bubbles up from below in the history of enclosure and conservation conveyed in the poem, "The Green Man." Owen-Fisher (2017, 20) writes how:

He loves that land, the flat, never ending stretch to the sea.
He can picture it in his mind''s eye, the wide expanse of nothing.
And everything.
Full of the feeling of home.

The creatures and trees "hold his muddy green heart inside" and "the things that grow in his flattened homeland, take root inside him." The green man is part and parcel of the Fens, and they are inside him. They are not a landscape; they are nothing and everything to him.

By contrast, in *Waterland* the flatness of the Fens is a business opportunity that makes "this nothing-landscape" ideal as "a drawing-board for the plans" of prosperous maltsters (Swift 2010b, 25). The Fens are a flat surface on which to draw the straight lines of their plans for improvement and a screen on which to project their phantasies of prosperity. Reclamation renders the sometime turbulent surface of water into the flat surface of land. Crick remarks that "the chief fact about the Fens is that they are reclaimed [from what? for what? for whom? one wonders], land that was once water, and which, even today, is not quite solid" (Swift 2010b, 16). Like Sayers' narrator, Crick wonders of land reclamation, "is it desirable, in the first place, which land should be reclaimed?" (17). The desirability of drainage is questionable indeed; in all probability and with the benefit of hindsight drainage was undesirable. For the non-Christian Crick, unlike the Christian Sayers, the Fens are also "a fairy tale land" and "a magical, a miraculous land" (11 and 121). Indeed, "the bare and empty Fens yield so readily to the imaginary—and the supernatural" (25) with "marsh-sprites and will-o-the-wisps" (26).

Unlike Di Palma, Swift is attentive to other pertinent aspects of the Fens, such as the theory of the humors based on the philosophy of the four elements, earth, air, fire, and water. The humors mix the elements with the mixing of water and earth producing the humor of phlegm (see Giblett 1996, 159–160). Fenlanders are phlegmatic people with a phlegmatic humor and a phlegmatic sense of humor. It goes with the place as if they have absorbed some of its characteristics. Crick describes how "phlegm [is] a muddy, silty humour" (Swift 2010b, 22) as "the Fens were formed by silt [. . .] which demolishes as it builds" (16). Phlegm, "or slime," is "an ambiguous substance. Neither liquid nor solid: a viscous semi-fluid [. . .] It deters the sanguine and the choleric and inclines towards melancholy" (341). Indeed, "melancholia and self-murder are not unknown in the Fens" (24), and "poor Fenland madmen and melancholics" (100) are not uncommon too. The mixing of water and earth also produces melancholy.

The phlegmatic is not stable in the periodic table of elements and humus. Even though the melancholic has traditionally been associated with the cold and dry, a visit to the cold and wet/dry marshes by the drylander can induce melancholy, whereas the indigenous wetlander by temperament is usually phlegmatic, but tends toward the melancholic. By temperament, the wetlander at home in the wetland is phlegmatic (associated with the watery), whereas the drylander visiting the wetland tends to be infected, or more precisely affected, by the melancholic (the earthy, though more precisely the earthy in combination with the watery).

In 2008 Swift (2010a, vi) wrote an introduction for the 25th anniversary edition of *Waterland.* In it, he acknowledges that initially in writing the novel "the Fenland setting, which could hardly seem more prosaic and monotonous" (even unpromising) was "a special, almost literal stage" on which the action and the lives of the character are played out and the arc of the story takes place. The Fens in the beginning of writing the novel "may have seemed to me the ideal nonsetting, the ideal flat, bare platform for my human drama" (vii), as if the Fens were the mere passive stage on which, and backdrop against which, the action of Swift's human drama took place. Swift goes on to relate that "what quickly happened [in the writing process] was that the apparent background became a foreground, *even a kind of principal character*" in the novel (vii; my emphasis). Wetlands, like the Fens, have agency, like characters in a human drama, like humans in the drama of real life.

Novels, invariably obsessed with human drama played out against the backdrop of a setting for the action, do not usually have a principal character who is not human, certainly not a wetland, such as the Fens, that figure in the nonhuman drama, the drama of nature and the elements. Novels can have a nonhuman principal character in the form of an animal, as in the famous case of the white whale Moby Dick in Herman Melville's novel of the same name, for many the greatest novel in English for the very reason it is not only about Captain Ahab's obsessive quest, but also about sea-faring, whales, whaling, whiteness, and nature. Besides the Fens themselves, *Waterland* has another nonhuman principal character in the form of eels "who feature prominently in the book" as Swift (2010a, v) says in his introduction with a chapter in the novel devoted to their natural and cultural history (Swift 2010b, 197–206). Having a habitat and wet landscape, such as the Fens, as a principal character is another matter entirely, at least until *Waterland* came along. Following in Swift's footsteps, Liam Davison's 1993 novel *Soundings*, a kind of Australian *Waterland*, set in the Koo-Wee-Rup swamp of south-eastern Australia outside Melbourne, and Peter Matthiessen's 1990s trilogy set in the Florida Everglades beginning with *Killing Mister Watson* are perhaps other cases in point of novels with a wetland as a principal character.[5]

In his 2008 introduction, Swift (2010, vi) also acknowledges that his "physical research for the novel was minimal." And it shows. He did not seem to have spent much time walking, punting, or sitting in the Fens absorbing local sounds, colors, and smells and writing about them in a journal as the basis and sourcebook for the novel, unlike Macfarlane's (2008) "Ghost Species," a recent instance of "new nature writing" about the Fens. *Waterland* is certainly not a book of nature writing about the Fens. Swift is not the Thoreau of the Fens. *Waterland* is not to the Fens as Thoreau's *Walden* is to Walden Pond and Woods in which they are not only principal characters, but also they, and their plants and animals, are minutely and lovingly described from first-hand observation and experience. Yet it is the physical aspect of the novel that meant Swift (2010a, vi) had to disabuse his Fenland readers that he is not a fellow "Fenny."

Although his physical research for the novel was minimal, in his 2008 introduction Swift (2010, vii) also goes on to acknowledge, "I did quite a lot of historical research," including without acknowledgement Charles Kingsley's (1873) "prose idyll," "The Fens." Swift's historical research shows too, as does Macfarlane's in his "Ghost Species." Macfarlane (2008, 111) also did some geographical and geological research as he describes the Fens as an "East Anglian [. . .] low-lying area of around 1,200 square miles," amounting to "one million acres" as Pryor (2019, xii) points out. Macfarlane then goes on to discuss the geology of the Fens with its surrounding limestone hills, chalk, and sand. Geology is literally the base for history and geography.

Like Di Palma (2014, 101–102), Swift (2010b, 18–19), Macfarlane (2008, 111–112), Pullman (2011, 99), Platt (2019, 137–138 and 141–143) and Pryor (2019, xv, 326–332) trace the history of the drainage of the Fens engineered by Cornelius Vermuyden and others in the seventeenth century. *Northern Lights,* the first part of Pullman's trilogy *His Dark Materials*, describes the Fens graphically as:

> that wide and never fully mapped wilderness of huge skies and endless marshland in eastern Anglia. The furthest fringe of it mingled indistinguishably with the creeks and tidal inlets of the shallow sea, and the other side of the sea mingled indistinguishably with Holland; and parts of the Fens had been drained and dyked by Hollanders [. . .] But parts of it had never been drained or planted or settled at all, and in the wildest central region [. . .] eels slithered and waterbirds flocked, [. . .] eerie marsh-fires flickered and waylurkers tempted careless travellers to their doom in the swamps and bogs. (Pullman 2011, 99)

No reasons are given for dyking and draining, and no consequences are drawn. They seem to be merely facts of history to note and pass over without comment.

Enclosure of the commons into private property produced highly fertile and arable land for farming as we have seen and as Macfarlane (2008, 112) notes when he describes how the draining of the Fens "revealed hundreds of thousands of acres of the most voluptuously rich soil in England"—enough to make farmers salivate and lust after it. Pryor (2019, xv and 342) concurs that, "once drained, fenland soils are very fertile," even "exceptionally fertile." Swift notes the unintended consequence that "reclaimed land shrinks" (Swift 2010b, 19). This history is the context 35 years later for Sylvia Tillyard's historical romance *The Great Level* (2018) that seem indebted too to Charles Kingsley's (1873) "prose idyll," "The Fens," that gives a wonderful potted history of the drainage of the Fens.

Tillyard's first-person narrator, Jan Brunt, is a Dutch engineer who is employed by Vermuyden in the drainage of the Fens in the seventeenth century that began in the reign of Charles I and was completed in the "Commonwealth" of Oliver Cromwell. Royals and republicans had a similar antipathy to wetlands and shared a capitalist drive to "improve" them by draining, converting them into dryland agriculture and deriving profit from them. The prologue to the novel is set on Manhattan Island in Dutch New Amsterdam (later New York) where Brunt has ended up and where he still pursues obsessively the Dutch compulsion "to attempt the separation of land and water wherever we find them muddled up" (Tillyard 2018, 7), and muddied together, one might add. The means to separate land and water from each other in what Macfarlane (2008, 112) calls the "terraqueous" (literally "earth-water") Fens is by what Brunt calls "embanking" whose aim is "to make an edge, a clear boundary and separation" between land and water (Tillyard 2018, 8). "Embanking" is his preferred euphemism for draining and reclaiming marshes and meres (see 96–97, 131–132 and 140). He prides himself that under his supervision "where once the land slid under the water in a disorderly way, the two are now divided by pilings that have given [the island] a firm edge all round its [southern] tip" colonised by the Dutch (8; see also p.43).

The Great Level reminds the reader that Manhattan Island was Manahatta, an island of marshes and that New York is a city of ghost marshes (see Giblett 2016a, chapter 9). Macfarlane (2008, 123) suggests that the Fens are a place of "ghost species," "a species that had been out-evolved by its environment, such that, while it continues to exist, it has little prospect of avoiding extinction. Ghosts endure only in what conservation scientists call 'nonviable populations'." This is a sad and sorry story of species' loss. The harsh terms of so-called conservation science belie the heart-ache and sense of loss experienced by Fenlanders. The ghost swamps and marshes of the Fens, and of the cities built on or in them, endure in what Walter Benjamin (1999, 462, 464) called a dialectical image in which "what has been within a particular epoch is always, simultaneously, what has been from time immemorial." "The primal phenomenon of history" of the swamp world is "present by virtue of its very oblivion."

The novel proper begins with Brunt meeting Vermuyden and being taken into his employ as an engineer in the drainage of the Fens. Vermuyden describes them as "a great wilderness [. . .] lying unimproved" (Tillyard 2018, 21). Brunt asks, "is it a lived wilderness, or at present water only?" to which Vermuyden replies, "it is inhabited, yes, Jan; but, the island of Ely excepted, by a lazy and barbarous people who trap eels and other such trash foods." Ironically, "Ely" means "island of the eels" (Macfarlane 2008, 112) so the name of the good Christian island is named after the trash food of the local barbarians! Brunt persists, "does not the place belong in some manner to these people, sir?" to which Vermuyden replies, "indeed not, or only by custom, for much rightly belongs, as underwater, to the realm, having formerly been the property of the King" (Tillyard 2018, 21–23). Customary belonging is the common law, the law of the commons. The custom of the commons meant the Fens belonged in common to all and were shared by all. This was the realm before it became crown land and before it was enclosed into private property. Vermuyden scoots over this history and legitimates "improvement" (and Pryor (2019, 342) follows suit) by way of drainage, embankment, enclosure of the commons and dispossession of Fenlanders. *The Great Level* traces in part this process, though it does not mention the commons, nor does Pryor (2019, 342).

Vermuyden's denigration of the Fens as being inhabited by barbarous people is reinforced by one tavern-goer who tells Brunt he has "heard it said that barbarous people live there." Another one adds "barbarous and godless, too, so I was told" (Tillyard 2018, 27). This is all second-hand hearsay that *The Great Level* contests as the Fenlanders are barbarous "it is said" (35), but they call all others uplanders, so the Fenlanders are "lowlanders," like the Dutch. Brunt relates later that “witches are said to be common amongst them, and many, I heard an old man say, still practise the pagan religion that long ago held sway across these islands” (49). In other words, the pagan religion is implicitly pre-Christian and explicitly barbarous, the religion of a barbarous people in a barbarous place. It is implicitly prior and opposed to Christian civilization and its churches as presented in *The Nine Tailors*. The Fens are also, like *Waterland*, a place of the supernatural with “marsh lights [and] will-o-the-wisps conjured by spirits” (119).

Brunt travels from Cambridge to Ely and is “heartened by the flatness of the country” as he travels “across a flat and marshy plain” and knows he is “going into the fenland” (30). He has “reached a world where water has its empire still” (31). Similarly, for Macfarlane (2008, 109) “entering the Fens always feels like crossing a border into another world.” What kind of world the Fens are Macfarlane does not specify other than pointing out different species of trees and sedges and implying the Fens are the world of the other, a wet world to the dry world, of a wetland people to dryland people.

For the Fenlanders in *The Great Level* the Fens are an expression of “the watery underworld” (Tillyard 2018, 122) with no pejorative overtones attaching to the concept of the underworld. The Fens are not hell, but heaven, albeit a watery world with watery human inhabitants who are described as amphibious for Macfarlane (2008, 111). This tends to make them sound like they were frogs, and indeed this was a pejorative jibe directed at them, but they were perfectly adapted to their watery habitat and home.

Brunt has come to wrest the watery world of the Fens from the empire of water by draining “the drowned country” and “the green world” (Tillyard 2018, 31) and thus making the wetland into dry land and so “redeem the drowned land” (54; see also p. 91). Drainage of wetlands and Christian redemption of the pagan went hand in hand. Indeed, for the Royalist William Dugdale (cited by Platt 2019, 15) in the seventeenth “works of Drayning are most antient [*sic*] and of divine institution.” Platt (2019, 15) comments that in “Christian mythology” God is “a kind of universal drainage engineer” who reverses the work of God as the universal wetland creator and who legitimates human drainage engineering.

Watery land can be drained, but “the watery air” (Tillyard 2018, 31) cannot be drained or dried. It proves a problem as a health hazard under the miasmatic theory of malaria, literally “bad air,” or ague that “the meres breed” from their “miasmic air” and that “haunt the ground” “like malevolent ghosts” (4, 102, 117, 172, 180 and 186; see also Pryor 2019, 4–5).

The flatness and wateriness of the Fens prompts Brunt to note that when the creator “created the earth, it was without form and covered all over with water, just as this land is in winter” (Tillyard 2018, 53). He is alluding to *Genesis* 1: 2 in the King James version of the Bible first published in 1611 before the events of *The Great Level*. The creator then divided the land from the water on the second day, as Brunt also notes (53). Brunt later draws the implication that "water was God's great love, the first thing, the dry land only a second thought" (76). Water covering the earth is the creator's (now named God) first and best work on the first day of creation. It was all downhill after that. Brunt does not draw the implication that drainage is fallen "Man's" last and worst work of doing God's work for "Him," and becoming God-like, arrogating to himself the position and function of God, like building a city to heaven (as in *Genesis* 11: 4). Wetlands are places that God forgot when "He" divided land and water. They hark back to his first and best work. After the Fall from grace, arrogant "Man" divides land and water from each other (and builds cities to the heavens in drained wetlands).

Like Swift's *Waterland, The Great Level* acknowledges the theory of the elements (and the humors in passing (192)) and the silty nature of the Fens. Brunt describes how, "in the marshes, where there might be a boundary [between land and water], but is none, the two elements are fused. The land is water-soaked and the water carries its heavy load of silt" (73). The job of drainage is to defuse the two elements of land and water by dewatering the land. Brunt goes on to state that water "must be drawn out from the land [. . .] and insinuated into a clean division of liquid and solid" (73). "Insinuated" implies something covert, surreptitious and underhanded as the dictionary confirms: "slide (oneself or a thing) slowly and smoothly into a particular place; manoeuvre oneself into (a favorable position) by subtle manipulation." Just as the land slid under water in the marshes of Manhattan and the Fens, so drainage slides the land out from under the water and so divides the land and water, solid and liquid. It brings the watery underworld out from below into the earthy over-world.

Brunt realises that "for one world to be made, another must die" (99). For the new world of dryland to be made, the old world of wetland must die in the capitalist logic of drainage engineering and private property. He has seen "that this unimproved world has its own way of being which will be lost" (99). For him this is a fact of life and of history, not a loss to be mourned. He is unlike his lover and protégé, the Fenlander Eliza (fenny, eely Eliza versus urban island Ely; barbarous, pagan Fens versus Christian church and civilization). She drains the swamp on her farm in Virginia (a long, romantic story). In the process, she has "to fell the cypresses which groaned as they came down into the swamps" and she is "pained by the loss" (209). She imbues the trees with sentience.

In *The Great Level* Eliza is a mediating figure between the ancient world of the Fens and the modern world of draining and reclaiming wetlands, such as the Fens of eastern England, the marshes of Manhattan and the swamps of Virginia. She has a foot in both worlds and tries to have the best of both, though in many ways they are antithetical to each other as a drained and undrained wetland is not possible. The Fens become the template for Brunt's "embanking" of Manhattan marshes and Eliza's draining of Virginian swamps. Yet whereas Brunt transfers the template unerringly, Eliza modifies it in more environmentally sensitive ways as she has "maintained several swamps away from the immediate surroundings of the house" (209). She wants to have her cake and eat it too.

The unimproved world of the wetland is its own world, complete and sufficient to itself. It is a way of being, an indigenous sacrality (see Webb 1996). It is not a social construction, nor a discourse, an institutionalised way of seeing, saying and doing, such as in Vermuyden's "Discourse Touching the Draining of the Great Fennes." The Fenlanders say "we have always been here" (121). They are people of the place. The Fenlanders belong to the Fens and they belong to them by "the right of custom" (121) in the customary law of the commons prior to and without a document of written law. The people belong to the land and the land belongs to the people, as in the Australian Aboriginal custom and concept of "Country" (see also Webb 1996).

The sacrality of the Fens is encapsulated by two figurines that are introduced in the "Prologue" when Brunt is in New Amsterdam (12) and that were given to him earlier in the Fens as related later (137). He carries them around in his pocket, polishes them, sits them on his desk during the day while he is writing and locks them up at night (137). They are antiquities that he has collected and displays (I am reminded of Freud and his desk) and fetishes that he worships as sacred objects. He relates that "they are women, goddesses or witches [. . .]. They are old women, many times mothers" (137). They are the Great Goddess, or Great Mother, of the swamp world (as discussed by J. J. Bachofen and catalogued by Marija Gimbutas; see Giblett 1996, 145–150, 196–198).

One figure is baked from clay, seeming to "symbolise" or convey the element of the earth, the other carved from sparkling rock, symbolising or conveying the element of water. In the clay figure "the gash of her belly button looks like an opening to the underworld" (12); indeed, it *is* an opening to the underworld of her swampy womb, and to the underworld of the womby swamp. Or more precisely and less coyishly, "her navel and the secret place between her legs" are described as "black as caves" (137). The slimy womb and womby wetland has black water, an uncanny abject of fascination and horror (see Giblett 1996, 31–34).

By contrast, "the other woman is fish-like and liquid [. . .] Legs fused like a mermaid's tail [. . .] She is a sea creature" (13). Brunt understands that "they belong together" (p137). They belong together as earth and water belong together in the marshy and swampy world and womb of the wetlands of the Fens. Separating them out completely is misguided as "without water, land is lifeless and barren. Wish then for the two together" (188). The womby wetland dries up when it is drained and cannot procreate new life. Land and water belong together, as they did in the Fens before "embanking" and as they did on the first day of creation where they were God's first and best work (as we saw in chapter 1). Long live the Fens!

NOTES

1. See also Bregman (2020, 309–315) for his account of the commons without referring to Marx, but invoking Elinor Ostrom (1990), critiqued by Standing (2019, 32) for "defining the commons in purely economic terms of 'extractability' and 'subtractability.' A commons can also be defined by heritage or customary use established over time that adds to our sense of belonging. The commons 'belongs' to the commoners, traditionally a group defined by customs." This is in line with Marx's and "Marxists" or "socialists", such as Thompson's and Williams', communist definitions of the commons. Conversely and equally, the commoners belong to the common in a mutual bond of belonging. For Ostrom (1990), however, there are no commoners, only "appropriators," another version of *Homo economicus*, whereas commoners are *Homo ecologicus*, earth householders. Ostrom's title and subtitle, *Governing the Commons: The Evolution of Institutions for Collective Action*, seems like a sequence of oxymorons. The commons precede and refuse governmentality and its institutions of collective action (including its discourses of institutionalised practices of meaning-making in disciplines), and promote communal use and action (for the genealogy of governmentality, see Foucault 1991). For a recent critique of Ostrom, see Obeng-Odoom (2021).

2. By ignoring this aspect of enclosure of the commons, Williams seems to have had a blind-spot about the specific history and geography of the Fens, even though he was living in them at the time (Williams 1973, 3), and even though he is the patron saint of eco-cultural studies (Giblett 2020, chapter 10). Perhaps he can be excused as he was following in the footsteps of Marx who does not mention the Fens either in his discussion of the enclosure of the commons into private property. All idols have feet of clay; all saints have blind spots.

3. Compare a similar and analogous contemporary situation in Oregon where a game "hunter" constructed his own private, artificial wetland on the banks of the Columbia River and planted it with rice to attract waterfowl so he and his "hunting" mates could take potshots at them and "harvest" them as they fly through the East Pacific flyway (Giblett 2014, 3–4), a latter day instance of "a duck decoy." These two instances of wetlandscape architecture in America and England could be called industrial "aquaorniculture" (from the Latin "aqua" for water and the Greek "orni" for bird) and contrasted with the constructing of wetland structures and the practicing of traditional aquaorniculture by Australian Aboriginal people using nets to snare waterbirds (as we will see in the following chapter).

4. "Swarming" is a loaded biblical term as swarming creatures come under the Levitical interdiction of being abominable (*Leviticus* 11: 41–43; see also Douglas

1966, 56). In mainstream western culture wetlands have not only been seen as the habitat of swarming creatures, such as alligators and crocodiles who transition from earth to water, but also as abominable themselves as they mix the elements of earth and water. With the exception of this comment, Williams had nothing to say about Clare's poetry about the Fens in his and his daughter's critical commentary in this edition, including in their discussion of Clare as "the finest naturalist in all English poetry" (Williams and Williams 1986, 213–217), earlier in their introduction to this edition (Williams and Williams 1986, 1–20) and in his otherwise exemplary discussion of Clare's life and work in *The Country and the City*, the pioneering work of ecocriticism and eco-cultural studies (Williams 1973, 132–141). By downplaying this aspect of Clare's poetry, Williams seems to have had a blind-spot about the significance of the Fens for Clare, similar to his blind-spot about the history and geography of the Fens (see n.2 above).

5. For further discussion of both novels, see Giblett (1996).

Chapter 4

Wetland Cultures of "Australia Felix"

From Mountain Ranges and Landscape Painting to Wetland Places in Environmental Artwork

Gariwerd, or "the Grampians," in Australia are largely known collectively as the mountainous region in western Victoria centered around the town of Halls Gap nestled in the valley between spectacular ranges and accessible from the east through a gap (hence the name) in the ranges.[1] Hardly surprisingly, these ranges "running in a north-south direction" and lying in "the path of rain-bearing westerly winds" (Wilkie 2020, 6) created magnificent wetlands with the water flowing off the ranges, collecting in basins in the ranges and out around their base in the surroundings. The Grampian mountain ranges are a watershed with water catchments flowing down to the east and west, feeding rivers flowing to the north and south, and creating surrounding wetlands and well-watered plains. The Grampian mountain ranges are "the head," or top, of these catchments with former wetlands there too.

The first Englishman to record the existence of Gariwerd and its wetlands was probably the explorer/surveyor Major Thomas Mitchell in 1836. Mitchell described the rugged mountains and blessed plains in terms of the European aesthetic conventions of the sublime, the beautiful and the pleasing pastoral picturesque. He dubbed western Victoria "Australia Felix," meaning "Australia the blessed" because of its well-watered grasslands. In his "Description of Australia Felix" he noted how "small rivers radiate from the Grampians" (Mitchell 1839). Not surprisingly, this area was the happy hunting ground for squatter-pastoralists as the well-watered mountains and blessed plains lent themselves readily to establishing sheep stations. It was also the happy hunting ground for the settler-colonial landscape painter Eugene von Guérard who depicted the rugged mountains and blessed plains in terms of the European aesthetic conventions of the sublime, the beautiful and the pleasing pastoral picturesque too. Pastoralism and landscape painting went hand-in-hand and were mutually reinforcing colonizing enterprises. The picture told the story; the story enacted the picture.

The Grampian mountain ranges are important geologically and biologically. The Heritage Council of Victoria (2005) states of their significance that:

> The Grampians comprise a complex of sandstone ranges rising abruptly from the Western Plain. From a distance they are most spectacular with their serried ranks of precipitous peaks rising to 1,164m at Mt. William. The area provides some of the most beautiful and diverse habitats for native flora in Victoria with over 1,000 species of ferns and flowering plants, many endemic to the area. A wide range of habitats resulting from the diverse topography, micro-climates and vegetation have provided secure refuges for many wildlife species including a number of rare species.

To what extent this "statement of significance" includes the wetlands surrounding the Grampians National Park is not clear (for further discussion of the geology and biology of the Grampians, see Wilkie 2020, 1–11).

Looking down from a plane flying from Melbourne to Perth the Grampians mountain ranges are certainly spectacular and their "serried ranks" are clearly visible, as are many of the magnificent Grampians wetlands, even when they are dry, such as Mt William (or Big) Swamp. Mountain ranges in traditional Chinese cultural geography are dragons. They are the backbones of the land. Without the spine of the land, the land would fall apart. The Grampians have four distinct ranges lying back to back or face to face with each other like four intertwined dragons. Water bodies in traditional Chinese cultural geography are places where the Tao, or the Way, begins, where new life springs. They are the internal organs of the land that hang off the spine. Without these organs, the land would die. They give and nourish life by supplying and purifying water. Traditional Australian Aboriginal cultures have a similar view of water bodies (Giblett 2020e).

The Grampians are generally not known for, or associated with, the wetlands that encircle the mountain ranges like a sapphire necklace. Sapphire for Pastoreau (2001, 7) is "a truly celestial color. Its blue, often compared to that of the sky, is said to have healing powers." Blue waters depicted in paintings, such as von Guérard's of swampy Fyans Creek near the Grampians (see figure 4.1), reflect the sky. All living waters reflecting the celestial blue sky are truly terrestrial and have life-giving and restorative powers. On maps of the Grampians National Park and the surrounding area, they are shown in blue in accordance with mapping conventions that show water bodies, or not at all.[2] Sapphires can also be black which is fitting for the mainstream denigratory, western cultural color-coding of swamps as

Figure 4.1 Eugene von Guérard, "Mount William from Mount Dryden," 1857 Oil on canvas, 61.5 x 91.5 cm. *Source*: Collection: Art Gallery of Western Australia

black waters, as blots on the landscape and in the ledgers of farmers and pastoralists.[3]

Benjamin Wilkie in *Gariwerd*, his otherwise largely exemplary and very useful recent environmental history of the Grampians, is so fixated on the mountains, the ranges, and the national park that he ignores, with a couple of exceptions, the larger bioregion and its wetlands. He might have discussed them, but does not, in relation to water, waterways, reservoirs, rivers and drainage systems (see, for example, Wilkie 2020, x, 7 and 85). Water is abstracted from the specific, living bodies of wetlands, and turned into a dead commodity to be transported, bought, and sold. This is a sad and sorry story which seem to be the wont of environmental historians, cultural geographers, and urban designers who neglect the life-giving properties of wetlands (as we will also see later in the present chapter and in later chapters).

The exceptions are when Wilkie acknowledges wetlands coincidentally, either in mapping Aboriginal country (though there is no key to this map; Wilkie 2020 25 figure 2.1), or in discussing George Augustus Robinson, the Protector of Aborigines, and his account of 1841 of Aboriginal peoples' uses of marshes (Wilkie 2020, 31–32). Wilkie (2020, 31) describes this usage as "aquaculture," but it is more precisely paludiculture (from the Latin word "palus," meaning "mire" or "marsh"). When Wilkie (2020, 48, figure 3.1) maps pastoral stations, he also shows the prominent mountains, major rivers, bigger towns (not Halls Gap), main roads, Lake Lonsdale, but no wetlands. It is as if pastoral stations wiped pre-contact wetlands off the face of the earth and wrote them out of the history and geography of the Grampians, which they did try to do, and Wilkie largely follows suit. It is as if wetlands are relics of a bygone Aboriginal era. Neo-colonialism is alive and well in 2020. Wetlands are still there!

Despite being a blot, or black spot on the landscape and in the ledgers of farmers and pastoralists, these wetlands played an important role in the lives of Aboriginal and European peoples. Weaving together the strands of local and environmental history with Aboriginal story of Gariwerd, and illustrated with landscape paintings by von Guérard and with environmental artwork by the local artist Carole Mules in mixed eco-dyed fabric collages creating textured renderings of the wetlands of western Victoria, the present chapter celebrates the life of the wetlands in and around the mountain ranges of the Grampians in western Victoria and their role in the lives of the people who have called this place home for many generations. The Grampians wetlands are found in the region bounded by the roads between the towns of Horsham to the north, Ararat and Stawell to the east and north-east, Willaura to the south-east, Dunkeld to the south, and Mooralla to the west. Based on archival and contemporary research into Grampians wetlands, this chapter conserves and conveys the natural and cultural heritage of the wetlands of the area for present and future generations.

ABORIGINAL SIGNIFICANCE

Aboriginal peoples have had long association with Gariwerd, the Grampian mountain ranges and its surrounding wetlands. Gib Wettenhall (1999, 6) in *The People of Gariwerd: The Grampians' Aboriginal Heritage* states that the Grampians have "immense importance as home and spiritual sanctuary to the two Aboriginal language group and their predecessors who had shared the ranges [and wetlands, I would suggest] for thousands of years." Their home was the bioregion of the Grampians catchment. Wettenhall (1999, 6) goes on to describe how "aboriginal people's territories were often catchment-based with rivers and mountain ranges acting as natural boundaries" and encompassing resourceful wetlands (as he goes on to discuss and as I will discuss later).

The *Grampians National Park Management Plan* begins by stating that:

> The Indigenous Nations have a long association with the Grampians–Gariwerd. The use of one site in the Victoria Range (Billawin Range) has been dated from 22 000 years ago. Recent archaeological investigations have demonstrated that there was intensive Aboriginal occupation of Grampians–Gariwerd. The

draining of Lake Wartook in 1997 for maintenance works exposed 32 sites around the margin of what was once a swampy basin. (Parks Victoria 2003)

The *Management Plan* is plagiarising Wettenhall (1999, 6 and 25), but neglects to go on to mention that "the 32 sites" around "the swampy basin" of "Lake" Wartook are now submerged beneath Wartook Reservoir. In other words, Wartook was once a wetland, or more precisely, a tarn, a perched, mountain wetland, though for Wilkie (2020, 85) it was merely "a natural basin." Wettenhall (1999, 25) does not neglect to go on to mention that "similar in distribution and content in other western Victoria wetlands [. . .] the Werdug ['Wartook' is a corruption of this Aboriginal name] complex bears all the hallmarks of village life, often established for months at a time." The established hallmarks and built structures of Aboriginal occupation and inhabitation were ignored by the early explorer, Major Mitchell who remarked that the territory was "in the state of nature," when in fact it was in the state of culture.

According to the regional water utility, "Lake Wartook is located on the MacKenzie River in the central Grampians. It is an important water resource in the region [. . .] The lake also provides important environmental flows" (GWM Water ND, online). For what or for whom these environmental flows are provided is not specified. It certainly provides continual environmental flows to MacKenzie Falls, the scene of spectacular thrills for tourists (discussed in greater detail below). GWM Water neglects to mention that MacKenzie River is seasonal, and so would the Falls be if it were not for the artificial "Lake" of Wartook Reservoir.

Despite the loss of these sites and this wetland, the Management Plan goes on to argue that "the park's Indigenous culture is rich, diverse, and living, but will need support and protection to be sustained" (Parks Victoria 2003, 22). Part of that support can come through providing knowledge and interpretation of the rich, diverse, and living culture and nature of Grampians wetlands (for further discussion of the Aboriginal significance of the Grampians, see Wettenhall 1999 and Wilkie 2020, 13–36).

EUROPEAN EXPLORATION

The first Europeans to visit the area noted its wetlands. Rod Bird (2014, 3) in his self-published history of wetlands in south-west Victoria writes that "at settlement by Europeans the land was full of swamps and lakes," yet it was hardly full of wetlands as it also had mountain ranges, such as the Grampians, that Europeans could easily see and note, and did so. Paul Carter (1987, 111) describes how the explorer/surveyor Major Thomas Mitchell "traversed the

Grampians regions during a very wet season; during a very dry season the same country wears a quite different appearance." Quite so, as much of it was made up of wetlands. In fact, the very wet season of 1836 made the region into what Wettenhall (1999, 6) calls "the maze of swamps." Wilkie (2020, 6) remarks that "the region was relatively wet just before the arrival of the Europeans in the late 1830s [. . .]. It was fateful that the Europeans would first encounter the region in a period of relatively high precipitation and moisture, and therefore lush, alluring vegetation."

On June 30, 1836, Mitchell (1839) climbed Pyramid Hill and later waxed lyrical in his journal that "I stood, the first European intruder on the sublime solitude of these verdant plains as yet untouched by flocks or herds." More specifically, the country "appeared to be undulated, open, and grassy" with forested land that "opened into grassy and level plains" and "open grassy plains, beautifully variegated with serpentine lines of wood" (Mitchell 1839). This country was, as Mitchell first indicated and as Wilkie (2020, 70) recently reiterates, "precisely the kind of landscape sought after by nineteenth-century pastoralists." Mitchell was an explorer/surveyor by profession or trade an appointment to and for pastoralists. One followed the other in lockstep with devastating impacts on indigenous peoples and places. In 1938 James Cowell (cited by Wilkie 2020, 69) wrote in *The Horsham Times* that "with man's sheep came destruction of this verdant growth," precisely the very object of desire that attracted them in the first place, and that they then destroyed. Some white men kill the things they supposedly love, like grasslands, and those they hate, like wetlands.

The Grampians wetlands were seasonal, as are many Australian wetlands, a situation often held against them by explorers and setters (see, for instance, George Webb"s denigration of Lake Monger in Perth for not being "like a lake at home" and "drying up" seasonally and inconsiderately according to Eurocentric criteria; Giblett 1996, 57–58). Yet the seasonal wetting and drying of Australian wetlands as noted by one of the earliest European observers of western Victoria in George Augustus Robinson, the "Protector" of Aborigines from 1839 to 1849, makes them highly productive and important as a food source for Aboriginal people (both discussed below).

In what Powell (1970, xxv) calls "an unusually wet winter—almost uncannily avoiding some of the least promising areas," Bird (2014, 3) describes how Mitchell "struggled across these landscapes," or more precisely wet landscapes, in south-west Victoria and named the area initially as "Eden" and later famously as "Australia Felix," "Australia the Blessed" (Wettenhall 1999, 34). Mitchell wrote that:

> We had at length discovered a country ready for the immediate reception of civilised man; and destined perhaps to become eventually a portion of a great

> empire. Unencumbered by too much wood, it yet possessed enough for all purposes; its soil was exuberant and its climate temperate; it was bounded on three sides by the ocean; and it was traversed by mighty rivers, and watered by streams innumerable. Of this Eden I was the first European to explore its mountains and streams, to behold its scenery, to investigate its geological character and, by my survey, to develop those natural advantages certain to become, at no distant date, of vast importance to a new people. (Mitchell 1839; cited by Carter 1987, 111 and 254–255)

Carter (1987, 255) concludes, "in short, Australia Felix was a picturesque country," or more precisely, some aspects of it were.

Carter overlooks that "Australia Felix" was also, in short and in part, a sublime country, or more precisely, some aspects were, as Mitchell goes on to write how:

> from the summit of Mount Abrupt I beheld a truly sublime scene; the whole of the mountains, quite clear of clouds, the grand outline of the more distant masses blended with the sky, and forming a blue and purple background for the numerous peaks of the range on which I stood, which consisted of sharp cones and perpendicular cliffs foreshortened so as to form one grand feature only of the extensive landscape, though composing a crescent nearly 30 miles in extent: this range being but a branch from the still more lofty masses of Mount William which crowned the whole. (Mitchell 1839; cited by Wilkie 2020, 44 and in part by Wettenhall 1999, 6)

The sublime in European landscape aesthetics has generally been concerned with vertiginous mountainous landscapes and Mitchell follows the convention. It is as if Mitchell were singing here from the song-sheet of the sublime penned by Edmund Burke and elaborated by Immanuel Kant in the eighteenth century (see Giblett 1996, chapter 2). The sublime is one of the major European aesthetic and cultural modalities that goes back to Longinus in the first century of the Common Era and comes forward to Lyotard in the twentieth century. For Lyotard (cited by Giblett 1996), "around the name of the sublime, modernity triumphed," not least over wetlands. The sublime is what could be called "a secular theology," a theology for a world without the Judeo-Christian God, or for a world in which "God is dead" (see also Giblett 1996, chapter 2).

The sublime was part of Mitchell's cultural baggage that he took with him on his expeditions in the nineteenth century and used in his descriptions of the "the sublime peaks" and "the sublime solitude of these verdant plains." Mitchell quite rightly associates the sublime with immensity in the vertical and horizontal planes (plains) and evokes the affect (bodily or visceral emotion) of solitude conjoining subject and object. The sublime is not a quality of the subject (the "I") alone, nor of the object (the "scene") solely, but of both together. The sublime is also a product of the position and stance of the subject ("I stood") in relation to geometrical objects (e.g., "sharp cones and perpendicular cliffs,") as famously depicted and evoked in Caspar David Friedrich's painting of a solitary "Wanderer above the Mist, or Mountaineer in a Misty Landscape," of *circa* 1818. In Mitchell's terms, this painting could be called "in sublime solitude in, or of, the mountains." It is as if Mitchell were transporting and translating the horizontal mid-ground in Friedrich's painting from grey mist to green plains, from image to word, from Germany to Australia only eighteen years later.

Carter (1987, 255) goes on to note that "in Mitchell's published narrative" the description quoted earlier about Eden "occurs in the entry for July 13, *before*, that is, Mitchell had entered Australia Felix." Mitchell, in fact, had called this area at the time "Eden," not "Australia Felix" at all; he had called it "Eden" when he had seen it and before he had entered it. "Eden" was the view of the country from afar as picturesque scenery in a pleasing pastoral prospect, whereas "Australia Felix" was his name for the region as productive pastoralist country close-up after he had entered it and when he later described it (as we will see shortly, and which Carter does not quote for some strange reason). The biblical gates of Eden are closed and guarded by an angel with a flaming sword for anyone wishing to re-enter it (*Genesis* 3: 24), while the portals of "Australia Felix" are open for any white male settler wanting to enter it (or her) guided by Mitchell with a flagrant pen—an exceptional case in which the pen *is* mightier than the sword.

Mitchell serves as a pander or pimp for white men entering "Australia Felix" when in his later description of, and introduction to, it he waxes lyrical that:

> The land is [. . .] open and available in its present state for all purposes of civilised man. We traversed it in two directions with heavy carts, meeting no other obstruction than the softness of the rich soil; and, in returning over flowery plains and green hills fanned by the breeze of early spring, I named this region Australia Felix, the better to distinguish it from the parched deserts of the interior country [. . .] flocks might be put out upon its hills or the plough at once set to work in the plains. (Mitchell 1839; cited in part by Bird 2014, 3)

In other words, "Australia Felix" is soft, flowery and virgin land that is open and available for "civilized" white men to penetrate and deflower it (or her). Australia Felix is also moist, green and fertile land, unlike the dry and barren desert of the interior of Australia. Australia Felix is ready and ripe for insemination and procreation by white men in industrial agriculture and pastoralism in which they are active and the land is a passive receptacle or incubator. Mitchell color-codes, genders, and sexualizes the land. Strangely Carter does not cite this particular description of "Australia Felix" in his extensive use and discussion of Mitchell's work in *The Road to Botany Bay,* nor does Michael Cathcart in *The Water Dreamers,* his environmental history of water in Australia (as we will see in chapter 6 of the present volume). Are they embarrassed by, or squeamish about, Mitchell's old-fashioned patriarchal sexual politics? They are fairly typical for his day, age, masculine gender, heterosexuality, and politics.

George Augustus Robinson, the Protector of Aborigines from 1839 to 1849, was an early dissenter from Major Mitchell's bucolic view of "Australia Felix" and "the Major's Eden," as Robinson puts it. As Carter (1987, 110) points out, "according to Robinson, most of what Mitchell called 'lakes' were never more than 'lagoons covered with thick reeds and grass',' an apt description of an Australian wetland. Carter goes on to point out that Robinson wrote that two "lakes" south of Mount Abrupt at the southern end of the Grampians:

> called by Mitchell lakes, had no doubt water in them [at the time]. But even then it must have been so shallow that had he examined them he would have seen the grass. They are now dry. (Cited by Carter 1987, 110)

In other words, lakes aren't shallow, don't have grass, and they don't dry up in Robinson's landscape lexicon. Robinson, like Webb, applies Eurocentric criteria for lakes to Australia. Mitchell for Robinson should have known better and not called these water bodies "lakes" when they didn't deserve them, or fit the criteria. Yet Robinson writes here another apt description of an Australian wetland that can be, and usually is, shallow, can have grass, and can and does dry up.

In July 1841 Robinson (1998, 306) was among the Grampians wetlands where he "passed several dikes dug by the natives for draining small lagoons into the large ones for the purpose of catching eels." The following day Robinson (1998, 308) was at "the confluence" of a creek and marsh where he:

> observed an immense piece of ground [of at least 15 acres] trenched and banked, resembling the work of civilized man[,] but which on inspection I found to be the work of the Aboriginal natives, purposely constructed for catching eels.

"Resembling" belies the fact that, indeed, it *was* the work of civilized people as Aboriginal people, in John Stuart Mill's definition of civilization, undertook "human improvement" and were "rich in the fruits of agriculture" (cited by Mayes 2018, 36), or aquaculture and paludiculture in this case. So much for the country being ready for the reception of civilization, as Mitchell stated previously as it was already civilized and had been for tens of thousands of years prior to his arrival there. So much for civilization beginning in Australia in the late eighteenth century, as Manning Clark famously claimed as the opening gambit of his monumental, multi-volume history of Australia (cited by Mayes 2018, 26).

Aboriginal people also demonstrated ownership of their property as they, in John Locke's terms, "mixed their labour with nature" (cited by Mayes 2018, 34). Yet contrary to Locke and his theory of property, rather than transforming the commons or wasteland into private property, their labor maintained the commons in common. Robinson (1998, 308) commented repeatedly that "these works must have been executed at great cost of labour" and "they evinced great perseverance and industry of the part of these Aborigines." "The plan or design of these ramifications" for Robinson (1998, 308) were "extremely perplexing" and he found "it difficult to commit to paper." He still tried to do so (Robinson 1898, figure 7.10). Robinson (1998, 307) had previously commented that "these people must labour to get precious food."

Not only does Mitchell boost "Australia Felix" as available for colonisation by industrial agriculture and pastoralism, but also by roads, towns, and "county divisions" along English lines:

> this territory, still for the most part in a state of nature, presents a fair blank sheet for any geographical arrangement whether of county divisions, lines of communication, or sites of towns etc. (Mitchell 1839)

This territory was, in fact, in a state of culture with villages (as we have seen), fish weirs, and eel traps (as we will see later). In accordance with the prevailing doctrines of terra nullius and tabula rasa, Mitchell equates Aboriginal people and their territory with nature, and thereby writes Aboriginal people, wetlands, and other specificities of their place out of history and geography. The traces of their lives and their cultures are wiped clean from the slate of the land (including wetland) in Mitchell's rendition of the landscape, the surface of the land presented to the eye. Aboriginal people were genocided by word, as they were soon in fact as many were massacred in this region, beginning with Mitchell's expedition, even though they were dependent on Aboriginal guidance and knowledge of the country (Koori Heritage Trust 1991; Wilkie 2020, 42, 62–63). The depths of the earth (including wetlands) are rendered into the surface of a blank sheet for the inscribing of roads, towns, and counties.[4]

The surface of the earth is also available for Mitchell as the basis for erecting monuments on it to the greater glory, power, and wealth of the British Empire. Mitchell (1839) concludes that such an enterprise "would be establishing a lasting monument of the beneficial influence of British power and colonisation." Such boosterism of the beneficent colonial master is the stock-in-trade of colonialism, when in fact it brought pain, disease, death, dispossession and destruction (as Mbembe 2019 remarks on many occasions). Erecting monuments on monstrous marshes, the grotesque lower earthly stratum, thus inscribing a place in space and a moment in time as a lasting memorial to itself, is also the stock-in-trade of patriarchal capitalist colonialism (see Giblett 2011, chapter 1; for further discussion of European exploration of the Grampians, see Wilkie 2020, 41–50).

LANDSCAPE PAINTING

Mitchell (1839; cited by Wilkie 2020, 70) wrote of his expedition that "we had been for some time travelling through forest land which now opened into grassy and level plains, variegated with belts and clumps of lofty trees giving to the whole the appearance of a park." Following literally and metaphorically in Mitchell's footsteps, Wettenhall (1999, 34) points out that "early settlers described the plains of western Victoria as looking 'just like a gentleman's park.'" One such early settler was John G. Robertson (Bride 1898, 34) who wrote in 1853 that "all the landscape looked like a park." Such a simile was the stock-in-trade of early explorers and settlers throughout Australia (Giblett 2011, 91–92). It belied the fact that, indeed, the landscape *was* a park, the product of Aboriginal firing and hunting practices in "the biggest estate on earth" (Gammage 2011; see also Giblett 2011, 93–94). The pastoralist's desideratum is the gentleman's park estate—on Aboriginal peoples' estate. The arriviste squatter could rise quickly up the social scale and class ladder by dispossessing Aboriginal people, acquiring their landed property, exploiting its resources cultivated by Aboriginal people and making a financial killing. Scottish settlers, who made up at least two-thirds of the pioneer settlers of western Victoria (Kiddle 1961, 14, 517 n1) and who had been cleared off estates in Scotland, cleared off Aboriginal people from their estates in Victoria (as we saw in the previous chapter; see also Wilkie 2020, 46).

To illustrate the simile of the country being like a gentleman's park estate, Wettenhall (1999, 34) reproduces von Guérard's painting, "Mount William from Mount Dryden" of 1857 and captions it as "depicting the park-like beauty of western Victoria" (figure 4.1). Indeed, this landscape painting does depict a park-like pastoral scene in the foreground with kangaroos grazing peacefully on grassland bathed in beneficent sunlight and ripe to be dispossessed by grazing sheep and cattle, but the mid-ground between the two mountains is unmistakeably swampy, with Mt William in the background. This swampy area is probably around Fyans Creek with Lake Fyans in the left background, another of the Grampians wetlands turned into a lake and a reservoir, the home of the Stawell Yacht Club and the source of drinking water for the town of Stawell and others.

Von Guérard's "Mount William from Mount Dryden" was bled-off (or cropped) and used as the cover illustration for the catalogue of the 1998 touring exhibition, *New Worlds from Old: 19th Century American and Australian Landscapes*, held at, among other places, the National Galleries of Australia and Victoria. Keaney (1998, 155) notes in her catalogue entry for this painting in the 1998 touring exhibition catalogue that contemporary newspapers "commented favourably on the beauty of the natural landscape [. . .,] commending von Guérard for his composition and the chosen vantage point. The artist was praised for his empathy with the Australian landscape." He was not praised for his empathy with the Australian wet landscape that makes a cameo appearance in this painting in the mid-ground and that appears centre stage in his painting of Mt William (discussed below in this chapter).

Contemporary newspaper reports also described von Guérard's painting "in terms of his national significance" (Keaney 1998, 155). More to the point, von Guérard was an Australian artist of nationalist significance for his paintings of the settler-colonial landscape that made the unhomely land homely by depicting it in terms of the European landscape aesthetic of the sublime, the beautiful and the picturesque, often in a single painting, as in "Mount William from Mount Dryden." Nothing much has changed to his status in over a century as he is still regarded as "arguably Australia's, and certainly [the state of] Victoria's most important colonial landscape painter," as McDowall (2014) puts it.

Keaney (1998, 155) relates how von Guérard's "Mount William from Mount Dryden" was acquired by the Art Gallery of Western Australia in 1972. She concludes that this painting "remains one of that gallery's most significant colonial Australian paintings." She does not give any reasons for this conclusion, but I hazard a guess that has this status because it contains many well-worn clichés of landscape painting in mainstream Western culture executed dexterously, such as the pleasing and picturesque pastoral foreground with native animals grazing peacefully ready to be supplanted by introduced farm animals, a wetland subsumed fleetingly in the mid-ground (as all wetlands should be fleeting according to mainstream western thinking about the landscape) providing a source of water for the settler, a sublime mountainous background topped by beneficent, rain-bearing clouds and overseen by a few raptors flying magisterially in the sky above the immensity and expansiveness of the land below as an emblem of God's providence overall.

Unlike von Guérard, who saw the land in terms of the sublime, the picturesque' and the beautiful, and Mitchell, who also saw the land in sexual, sublime, and picturesque terms, Mitchell's second-in-command, Assistant Surveyor Grenville Chetwynd Stapylton, saw the flat or undulating land in similar picturesque and pastoral terms to both men when he "came across 'a vast grassy plain'" that he described as "the most romantic view that can be imagined" (cited by Wilkie 2020, 44). He also saw the Grampians as "a beautiful and romantic appearance" (cited by Wilkie 2020, 44), unlike Mitchell (1839) who saw them as "the sublime peaks" (as we have seen). Stapylton did not differentiate the two different landscapes of the flat or undulating plains and the rocky or steep mountains from each other in terms of two different modes of aesthetics of the sublime and the picturesque as Mitchell and von Guérard did.

Stapylton also saw the land in much more practical ways than Mitchell did as he, in Bird's (2014, 3) words, "had the task of coping with this soft, exceedingly swampy country. His view of the landscape was not so lyrical for he was responsible for the extrication of bogged wagons and boat carriages and the management of unruly convicts." Soft and boggy wetlands do not lend themselves to being easily traversed by heavy boat carriages and baggage wagons. Wetlands, at least since Roman times in Britain, have been denigrated for the transportation difficulties they pose to the imperial coloniser (Giblett 1996, 18–19, 206–207). Looking at the land from the point of view of transportation (as the assistant surveyor Stapylton did) is much different from looking at it from that of its potential for industrial agriculture and pastoralism (as the explorer/surveyor Mitchell did) and the aesthetics of the pleasing prospect of the picturesque pastoral and the rugged mountainous sublime (as Mitchell also did and as the landscape painter von Guérard did too).

Bird goes on to describe how south-west Victoria was "a soft land" and to quote from Dennis Conley's contribution to *The Western Plains: A Natural and Social History* published by the Australian Institute of Agricultural Science in Victoria in 1986:

> the untroddden sward [. . .] was literally comparable to a bed of sponge; our horses sank to their fetlocks with every step [. . .] A two years' occupation in most instances rendered a station so "firm" that horse racing, kangaroo, emu and dingo hunting [. . .] formed one of the principal sources of amusement to the light-hearted settlers. (cited by Bird 2014, 3)

The sward, the upper layer of soil covered by short grass and so desirable pastoralist land, was trodden by Aboriginal people for tens of thousands of years before the arrival of Europeans. The transformation of soft and spongy wetland into firm dry landmarks the sad and sorry story of "occupation," dispossession, and colonization by the armies of settlers with their cavalry of mounted hunters, stockmen and drovers, and their infantry of cattle and sheep. It marks the transformation of the native quaking zone of wetlands where the earth trembles and where Aboriginal people tread lightly into the feral quaking zone of drylands where introduced animals tread heavily and native animals are hunted and tremble in fear.[5]

MOUNT WILLIAM SWAMP

"The Wartook Reservoir," as Wettenhall (1999, 25) prefers to call it, was not alone in submerging an Aboriginal village and a Grampians wetland. He also goes on to state that "archaeologists now consider that other wetlands, such as the Moora Moora Reservoir [also in Grampians National Park . . .], could well exhibit similar signs of intensive occupation by Djab wurrung or the Jardwadjali" peoples (Wettenhall 1999, 25). So much for terra nullius. These signs were also evident much farther away in what Wettenhall (1999, 24) calls "the Mt William wetlands" with the construction of fish weirs and the use of eel traps (as noted and illustrated by George Augustus Robinson in his journal of 1841; see Presland 1977; Wettenhall 1999, 24; Wilkie 2020, 31–32). They practiced both aquaculture and paludiculture, the cultivation of wetlands, also observed by Robinson (cited by Bird 2014, 4) when he noted that "the great swamp abounds in rushes, the roots of which are edible and afford the natives an ample supply." Wettenhall (1999, 34) describes how, "on the plains [of western Victoria], large numbers of lakes and swamps offered a wide range of reliable food resources, both plant and animal." The fact that Aboriginal people regulated the flows of water in wetlands and rivers with weirs and channels, not only for catching fish and eels, but also for cultivating edible roots and rushes and attracting waterbirds for their consumption suggests that they were practitioners of paludiculture. Aboriginal people also netted waterbirds as noted by Major Mitchell (1839) in his journal when he observed them constructing wetland structures and practicing what could be called traditional "aquaorniculture" (from the Latin "aqua" for water and the Greek "orni" for bird).

Looking at the land from the point of view of food resources (as Aboriginal people did) is different from that of the landscape painter (as Eugene von Guérard did). Swamps and marshes may be viewed in the mid-ground of the picturesque pleasing prospects as in von Guérard's "Mt William from Mt Dryden," or viewed in the foreground as in "Mt William and part of the Grampians," or viewed from a distance in the background as in "Tower Hill" of 1855 (National Gallery of Victoria, n.d.). The latter painting depicts in the mid-ground what Andrew Sayers (1998, 154) describes in *The New Worlds from Old* exhibition catalogue as "the shallow marsh [. . .,] part of a connected series of wetlands and a fertile breeding ground for birds and other wetland animals." The wetland is a water-filled, dormant volcanic crater now in the Tower Hill Wildlife Reserve in the Warrnambool area of south-western Victoria.

Ten years later (figure 4.2) von Guérard painted a painting of another wetland in Western Victoria based on a sketch he had done about ten years before while he was traveling and staying in the region (Pullin 2011, 126–127; see also Pullin 2018, 179 and map unp.). According to one writer on this painting, "although this scene was on Mount William Station, von Guérard chose to paint it as a primeval landscape." In other words, although the scene was part and parcel of pastoral land use, von Guérard chose to paint it as a primeval hydrological wet landscape of marsh and slime in the fore- and mid-ground, and as a primeval geological dry landscape of rock and the sublime in the background. The horizontal, supine wet landscape is countered by the vertical, erect dry landscape in the background in a classic, masculinist gendering of the land. Mountainous clouds surmount the scene and shroud the top of Mt. William as an emblem of God's overarching omnipotence, if not beneficence.

Figure 4.2 Eugene von Guérard, "Mount William and Part of the Grampians in West Victoria," 1865 Oil on Cardboard 30.3 x 40.6 cm. *Source*: National Gallery of Victoria, Melbourne Collier Bequest 1955 (1562–5)

It is as if von Guérard set out in this painting to illustrate Mitchell's description of "the sublime scene" from the summit of Mt Abrupt cited earlier in which he was singing from the song sheet of the sublime. Von Guérard's painting with its strong horizon in the mid-ground and with placid water and marshy vegetation in the foreground could also be called a depiction and evocation of the slime in the horizontal of wet landscapes, as distinct from the sublime in the horizontal of the verdant plains (as Mitchell did as we have seen) or of desert landscapes (as identified and discussed by Rudolf Otto 1950, 69). Sheep and cattle are *not* grazing safely in this pastoral landscape and no damage by their hard hooves is evident in the exposed foreground. Rather, native waterbirds wade and browse placidly in the water or fly smoothly above the water, emus congregate convivially in the marshy vegetation, and a lone pelican floats magisterially in the air above. In the sense of being a benign, prelapsarian landscape of God's good earthly creation, the scene depicted is pastoral; in the sense of being a scene of chaos, it is also a primeval wet landscape before the fall into evil, into wetland drainage and into agricultural and pastoral damage.

This painting was not included in the *New Worlds from Old* exhibition (and so not in the catalogue either) for some strange reason, especially as this touring exhibition was held for several months in the National Gallery of Victoria where this painting is housed. Perhaps it was not included because von Guérard expressed too much empathy in this painting for the slimy swamps of the Australian wet landscape, and/or for playing around with the conventions of the European landscape aesthetic in the s(ub)lime (slime is the secret of the sublime, as Zoë Sofoulis construes it in her parenthetical portmanteau; see Giblett 1996, chapter 2) and/or for not singing from the same song sheet about the grandeur of nature, and so on, as most of the paintings in this exhibition do from which there were some other noteworthy absences (for further discussion of these absences and of the s(ub)lime in Australian wetlandscape painting and photography, see Giblett 2020a).

The wetland on Mount William Station in von Guérard's painting was called (unoriginally) "Mount William Swamp." The scene depicted in this painting is also a primeval wetlandscape before the fall into evil and land abuse. The "untrodden sward" and primeval slime in von Guérard's painting of swampy Fyans Creek in flood in his painting of "Mount William from Mount Dryden" (figure 4.1) becomes the devastated and desolate pre-industrial agricultural landscape of a blighted and drought-stricken farm in Arthur Boyd's painting "Irrigation Lake, Wimmera" painted in 1950 depicting the other side of the Grampians, the dark side of the pastoral wet daydream, a drought-stricken nightmare, or day-mare. This painting depicts a desolated wet landscape and tree graveyard on a farm in the Mallee/Wimmera region of western Victoria west of the Grampians with trees in a dam dying from salination or inundation or both. The lake depicted in this painting looks like it is well on its way to becoming irredeemably saline caused by the clearing of native trees and stripping of the topsoil for a reproduction of this painting, see (Giblett 2020a, figure 3.3, 58). Perhaps this painting is the visual expression of the dark side of Mitchell's verbal explorer-settler water dreaming, "Australia Execratus," "Australia the Cursed," to Mitchell's "Australia Felix."

The living wetland is transformed into a dying wetlandscape, a landscape painting of it, and a stark portrait of a once living being. This wetland that was once living is now a ghost swamp. Land is transformed into landscape and the face of the land ("land-face") into portrait. As land is to landscape, so face is to portrait, as naked is to nude (see Giblett 2009, 177–178). In Boyd's painting the ruined wetland and dry desert is desublimated into dried paint on the dry surface of the painting. The life of the land and the depths of the wet-land are surfacted into painterly surface. Soft, wet paint is applied to a hard, dry surface. Soft, liquid paint dries hard into solid painting. In von Guérard's painting of Mount William (or Big) Swamp, solid mountain is sublimated into the ethereal sublime while liquid swamp is surfacted into aquaterrestrial slime. The two paintings are a diptych presenting in the now, a dialectical image of a moment in the past and in the present of absent wetlands, "Australia Felix" and "Australia Execratus."

Eugene von Guérard's swamp still survives, or at least the middle of it is protected, perhaps because it is too boggy for farming and only good for water-birds. Bird (2014, 18) describes how "some 635 ha of unfenced Crown Land remains in the centre of the 1900 ha Mt William Swamp, but is not accessible to the public." This area is named as "The Big Swamp WR" on the accompanying map (Bird 2014, 18). "WR" is short for "Wildlife Reserve" (not "Refuge"), or previously "Management of Wildlife Purps," in other words, a hunting preserve, as in a survey map of 1880 (figure 4.3):

Mitchell's lines of communication and counties are writ large here in a rectilinear grid of lots on the surface of the earth. The fluid outlines of the swamp are subsumed beneath the grid, the life of the Great Goddess/Mother (or Mater) of swamps buried beneath a mathematised matrix ruled by set square and tee square. From matter to matrix marks the sad and sorry fate of many wetlands (see figure 2.1). The commons of Aboriginal land were enclosed and "alienated" into private property. Bird (2014, 18) captions a photo of "Mt William Swamp as it was in August 2011—nearly full for the first time in 65 years. Most of the area has been alienated and sown to pasture (note the fence across the lake)." The fence is another line, another inscription on the surface of the (wet)land (for further discussion of colonization of the Grampians and pastoralist dispossession of Aboriginal people, see Wilkie 2020, 59–76).

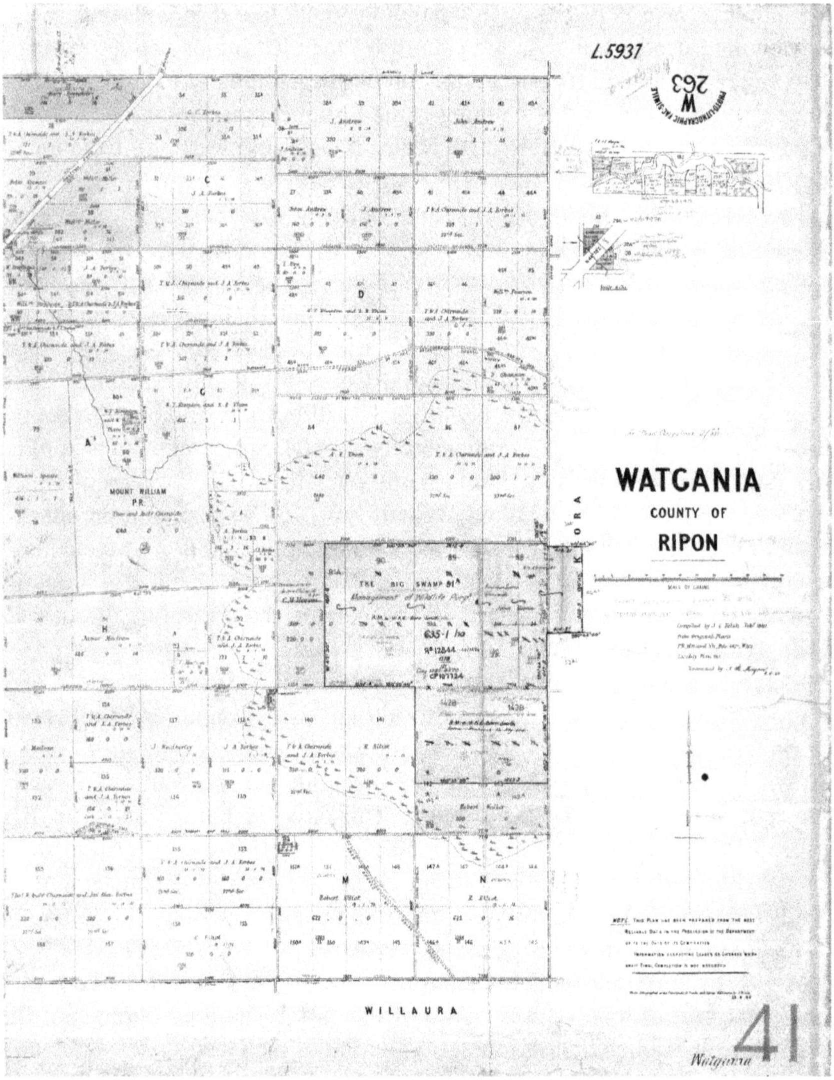

Figure 4.3 Watgania Parish, County of Ripon, 1880. *Source*: Courtesy of Carole Mules

OTHER GRAMPIAN WETLANDS

Other Grampian wetlands experienced a similar fate to Mt William Swamp. Bird (2014, 14) relates how "Bradys Swamp was once a magnificent wetland, filled by fresh water from the Wannon River, flowing out of the park from its alluvial fan. It was host to hundreds of Magpie Geese, Ibis, Brolga, and other waterbirds." The Grampian wetlands are noted as Brolga habitat. The water flow was reversed with devastating consequences. Bird (2014, 2) describes how:

> Heifer Swamp to the east was drained in about 1900 and the water was directed via a deep channel through a lunette into Bradys Swamp. In the late 1940s a shallow drain was dug across the swamp to create a new outlet into the upper Wannon River. As a consequence of the drain, Bradys Swamp is usually dry by January or February. Works to restore this picturesque wetland are required.

Even if it were not picturesque, it should still be restored. Being picturesque is not the only reason for restoration. Being water-bird habitat is a reason for protecting it.

As it is for another Grampian wetland in Lake Muirhead, close to Mt William Swamp. Rod Bird (2014, 20) describes how:

> This wetland, comprising both shallow and deep freshwater marsh, is a prime flocking site for Brolga in Victoria, with several hundred birds sometimes seen there.
>
> At least parts of the lake dry out annually and the birds seek crickets and other insects on the flats. In summer and autumn Brolga spend much of their time in flocks on stubble areas in the Willaura area [. . .] Lake Muirhead is also prime habitat for rare species such as the Freckled Duck. The south-west section has an area of flats that are subject to inundation. Parts are generally covered with reed and rushes while the rest is mud flats when not under water.

This temporal and spatial variation is fairly typical of Australian wetlands.

A much more positive story of a Grampians wetland is told recently with Walker Swamp. In June 2019:

> The landmark purchase of the 192-hectare Walker Swamp Reserve has been celebrated by partners and the community as a major step towards protecting the critically endangered wetlands in the Grampians. The reserve will permanently protect this important link in the large wetland system on the Wannon River Floodplain [. . .] Work to date to protect the Grampians wetlands includes: [. . .] 430 hectares of wetlands protected by landholders through the Wetland Defence Program and six one-day Farmplan21 workshops have been held, with 12 landholders attending. The workshops gave farmers the knowledge and skills needed to make sustainable land management decisions [. . .] The Eastern Maar Aboriginal Corporation and Martang Pty Ltd are involved in the ongoing cultural management of the wetlands of the Greater Grampians. A series of "Water and Country" events involving Traditional Owners included

> tours of the Budj Bim landscape, a children's school holiday program and an Aboriginal art competition and exhibition. The events promoted the unique wetlands and their biodiversity and cultural values to the community. (DELWP 2019, online; for Budj Bim see, Gunditjmara with Wettenhall 2010, 7, 13–16, 67)

The wetland cultures of the Grampians wetlands are finally being celebrated.

NATIONAL PARK

Grampians National Park was "set aside" in 1984 as an area of outstanding natural beauty and sublimity with mountain ranges and waterfalls, such as Kalymna and MacKenzie Falls, in accordance with the European landscape aesthetic reproduced in the American template for National Parks. Using the presence of a waterfall as an essential criterion for a national park is part of the modern tradition of aestheticizing nature in terms of what Dean MacCannell (cited by Giblett 2011, 159) calls "outstanding features of the landscape" (including not just a waterfall, but "a large waterfall"). It is also part of what he goes on to call "the modern touristic version of nature" which treats it as "a common source of thrills." Waterfalls, as Paul Shepard (cited by Giblett 2011, 159) points out, "have been primary tourist attractions for a thousand years"; national parks, usually with waterfalls, have been primary tourist attractions for a hundred years. Yet national parks are usually more than just tourist attractions. Or more precisely, the touristic functions of national parks are tied up with other colonialist, capitalist, and nationalist agendas and only comparatively recently with conservationist ones (for further discussion, see Giblett 2011, chapter 8).

The Grampians were, in the words of the *National Park Management Plan* (Parks Victoria 2003, 25), "generally unsuitable for farming," while Mitchell eulogized the surrounding fertile and well-watered plains as highly suitable. The unsuitability of the Grampians for farming conforms to the American template of "worthless lands" that was one of the impetuses for the setting aside of national parks in the United States. In the United States, as Alfred Runte (cited by Giblett 2011, 166) argues:

> National parks, however spectacular from the standpoint of their topography, actually encompassed only those features considered valueless for lumbering, mining, grazing or agriculture [what has come to be called the "worthless lands" thesis]. Indeed, throughout the history of the national park idea, the concept of useless scenery has virtually determined which landmarks the nation would protect as well as how it would protect them [. . .] not until the 1930s would

> wilderness preservation be recognized as a primary justification for establishing national parks, at least in the eyes of [the US] Congress.

The inclusion of wildlife preservation in the United States came as late as 1934 with the creation of the Everglades National Park in Florida. Perhaps it is hardly surprising that the preservation of wildlife should be brought about in relation to that most useless and worthless of areas for industrial lumbering, mining, grazing and agriculture, and to that least aesthetically pleasing of landforms, a wetland (see Giblett, 1996, chapter 1). In the Grampians, the wetlands surrounding the mountain ranges and later National Park are largely in private hands, with the exception of the middle of "Big Swamp" (as we have seen). These wetlands had some worth and usefulness as hunting preserves, sources of food and water for stock, and summer refugia for stock.

The Grampians National Park also followed the American template as the site for what the Management Plan (Parks Victoria 20013, 36) calls "scenic driving." The car did more than conservationists to create national parks in the United States and in Australia. Stephen T. Mather, the first director of the US National Park Service [established in 1916] recognized that park development was linked intimately to the growth of tourism, so he energetically built a second "pragmatic alliance [. . .] between the NPS and automobile interests throughout the country" (cited by Giblett 2011, 160). The touristic functions of national parks tied up with other colonialist, capitalist and nationalist agendas were worked out in the use of modern transportation technologies, such as the car. The car created a more dynamic set of picturesque or pleasing prospects with various stops for taking in the view. The car driver and passenger looked out front and side windows at sweeping panoramic vistas surrounding the car. Wetlands do not generally lend themselves to this point of view (for further discussion of the development of Grampians National Park, see Wilkie 2020, 102–112).

ENVIRONMENTAL ARTWORK

Wetlands do not generally lend themselves to a sweeping panorama in landscape painting, or to any sort of representation in landscape painting for that matter that uses conventions of the European landscape aesthetic of the sublime, beautiful and picturesque. Eugene von Guérard's paintings of the Grampians with a swamp in the foreground or the midground are the exception that largely proves the rule (as we saw previously in the present chapter). Wetlands are much more suited to the close-up point of view in the mixed media of environmental artworks, such as that by Carole Mules, of

Grampians wetlands. Mules depicts (they are not landscapes) these wetlands in fabric and stitching with flowing lines of waters and clumps of vegetation (figure 4.4).

Located near the Willaura Golf Course and south-east of the Grampians National Park, the Cockajemmy Lakes are what Mitchell (1839) called a "chain of lagoons" that leads from Gariwerd to the Hopkins River, feeds it and expresses above ground level the subsurface groundwater flows underground. "A chain of ponds" or "a chain of lagoons" were noted by many early European explorers of northern and southern Australia, including Mitchell in both areas (Massy 2020, 130; Mitchell 1839).

Mitchell (1839) named the Cockajemmy Lakes on September 20, 1836, and noted their hydrogeology:

> I observed that the lakes occurred at intervals in a valley apparently falling from the westward in which no stream appeared, although it was shut in by well escarped rocky banks. [. . .] No connection existed by means of any channel between them although they formed together a chain of lagoons in the bed of a deep and well-defined valley. On the contrary, the soil was particularly solid and firm between them, and the margin of the most eastern of these lakes was

Figure 4.4 Cockajemmy Lakes. *Source*: Courtesy of Carole Mules

separated by a high bank from the bed of another valley where a running stream of pure water flowed over a broad and swampy bed fifteen feet higher than the adjacent valley containing the stagnant salt lakes.

Wetlands aren't just what is visible on the surface, but are also manifestations and surface expressions of underground flows. Water isn't just a resource that can be seen, transported and commodified, but the life-blood of the body of the earth that flows through its internal, unseen veins and arteries.

Wetlands on the surface can dry up and disappear underground only to re-appear later in a wet year, as depicted in this artwork by Mules (figure 4.5). Mules reprises and reconstructs Eugene von Guérard's two paintings of Mount William (figures 4.1 and 4.2) with their abundance of water in the foreground or midground. In Mules's depiction the land is dry and parched with the Cockajemmy Lakes in the foreground and with abundant vegetation in the midground and background as a sign of underground water. The (wet) land has become textural, rather than primarily visual. It has ceased to have the smooth surface of water as in von Guérard's paintings, and has become the rough surface of cracked dry land. The wetland has disappeared in a dry year. The wetland may re-appear in a wet year. The wetland in Mules' artwork is reminiscent of Arthur Boyd's paintings of drought-stricken

Figure 4.5 Mt William from Willaura-Glenthompson Road. *Source*: Courtesy of Carole Mules

waterholes (Giblett 2020a, figure 3.3, 58; figure 3.4, 59) and Sydney Nolan's "Carcase in Swamp," yet with the hope and possibility that the wetland will reappear as seen in von Guérard's paintings.

Carole Mules depicts other Gariwerd wetlands, such as Lake Bunninjon shaped as an embryo, in mixed fabric (figure 4.6). Two human figures gather at, or in, the water, perhaps to share it and its bounty. Or two guardian angels of the wetland stand guard on either side to protect its waters, its plants, and animals. Or two ghostly virtual figures from the past. This artwork alludes to and plays intertextually with the biblical story of the Samaritan woman at the well. In the New Testament gospel of Saint John the story begins by relating how Jesus stops by Jacob's well, sacred to both Jews and Samaritans. Jacob was a Hebrew patriarch of the pastoralist type so he is a fitting figure to invoke in relation to the "Australia Felix" of western Victoria and Mules' artwork (*John* 4: 6–26). A conservation counter-theological rewriting of the biblical story in dialogue with Mules' artwork would go something like this.

Not so long ago an Aboriginal woman and a white pastoralist accidentally arrive at this wetland at the same time. She was dressed in a long white flowing robe and he was wearing a full-length brown moleskin coat, the standard uniform of the western districts' pastoralist. Both had come to this place for water, the woman to draw water for herself and others, the pastoralist to get

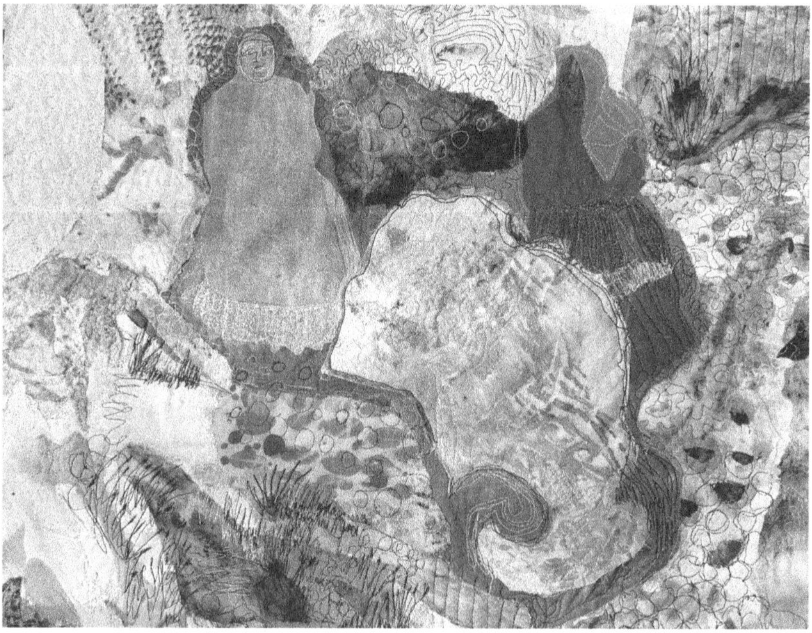

Figure 4.6 Lake Bunninjon. *Source*: Courtesy of Carole Mules

a drink of water for himself and to water his horse. The disciples had gone looking for food. The pastoralist was too lazy or arrogant to get down off his horse. Besides, he did not want to get mud and slime on his shiny boots, so he said to her, "Will you give me a drink?" The woman replied to him, "You are a white man and I am an Aboriginal woman. How can you ask me for a drink?" (for white people do not associate with Aboriginal people as a rule). The man answered her, "If you knew that I own this land and water, you would give me a drink." The woman replied, "My people own this land in common and this body of water. Everyone who drinks this water will be thirsty again, but whoever drinks the living water that our ancestors give them will never be thirsty. Indeed, the water they give them will become in them a spring of water welling up to eternal life." The man got off his high horse, came down to the wetland (as depicted in Mules' artwork) and said, "Where is this water? Where can I get some?"

The woman went on, "The time has come, the time is now, when all human beings should be worshipping neither God the father (the patriarch), nor (on) the mountain, nor (in) the city (eternal or temporal, celestial or terrestrial, Jerusalem or Rome), but worshipping the god who is breath, or air, or oxygen, the spirit that is overall and in all life, and most matter for that matter. All living creatures need the breath of life to live and need living waters to live too. The waters of wetlands are living waters welling up in the eternal now giving birth to life. Both air and water are sacred gifts that should not be polluted, but kept clean, owned in common and shared by all. The messiah and her disciples, the twelve apostles of conservation, explain this to us." The pastoralist had never heard of them so he asked, "Who is this 'messiah'?" Then the woman declared, "I, the one speaking to you—I am she." Just then the twelve environmental apostles showed up right on cue with food that they had gathered. The pastoralist could now see who they were as they introduced themselves to him. What a motley bunch of men and women they were. He declined their invitation to join them for lunch of bush tucker and rode back to his station mansion for his lunch of roast mutton and potatoes washed down with a glass or two or three of western districts' shiraz.

In the Eastern Orthodox and Eastern Catholic traditions, the Samaritan woman at the well is venerated as a saint, Saint Photine (literally, "light"), also transliterated as Fotini. She is the bearer of light, even the patron saint of light. She is a goddess of light as wisdom. She is also the goddess of living water in the well and elsewhere, such as wetlands as in Mules' artwork. She brings the light into the white pastoralist's heart of darkness and his benighted ideology of colonialism and private property. She brings him "out of the dark night" of colonialism (Mbembe 2021) into the light of day. She explains this to him and to us. She is a prophet who proclaims prophecy in poetic language. She is the leader of the twelve environmental apostles who explain her message to us too (Giblett 2020d, 35). She is the messiah of conservation who has come and is coming again and again. She is depicted with Jesus in many paintings engaged in what is called "The Water of Life Discourse" in which he delivers his institutionalised theological sermon to her as his sole recipient in a one-person congregation. Yet rather than a monologue, the biblical story is a dialogue. She is a prophet who proclaims prophecy in impassioned speech in dialogue with her listeners. Mules' environmental artwork depicts two equal figures engaged in the material and spiritual "water of life" dialogue around a wetland. Living waters matter!

NOTES

1. For the history of a Scottish/Latin name of a portion of the Scottish Highlands being applied to mountain ranges in western Victoria in Australia, see Wilkie (2020, 45).
2. For a map showing some Grampians wetlands, such as Brady's Swamp, Bryan Swamp and Lake Muirhead, and other water bodies, all depicted in blue, and not showing some Grampians wetlands, such as Mt William (or Big) Swamp, Walker Swamp and others, see Parks Victoria (2018). For a discussion of the mapping convention of depicting wetlands and other water bodies in blue, see chapter 7 of the present volume.
3. For black water as a blot both on the landscape of the pleasing pastoral prospect and in the English landed gentleman's ledger, see Giblett (1996, chapter 1) drawing on Wilkie Collins's *The Woman in White*.
4. For writing as inscription and as an instrument of power as distinct from writing as trace as a performance of inhabitation, see Giblett (1996, chapter 3).
5. For native and feral quaking zones, see Giblett (2009, chapter 1).

Chapter 5

Wetland Cultures of "Western Australia Felix"

From Mountain Range and Landscape Aesthetics to Wetland Wombs in Environmental Artwork

Like Gariwerd, the Grampians mountain range in western Victoria, the Stirling mountain range in Western Australia is not only a valuable recreational and tourist site for bushwalking, mountain-climbing, wildflower-watching, scenic driving, and sight-seeing, but also a vital place for wetlands, including those that John Charles Ryan (email to Giblett, May 30, 2020) evocatively calls "seasonal soppy meadows between the peaks," as well as those similarly seasonal wetlands created by run-off from the range in the surrounding plains. Like Gariwerd, the mountains of the Stirling Range have been privileged over its wetlands in keeping with the conventions of the European landscape aesthetic of the sublime theorised in relation to mountains by John Ruskin and practiced and popularized in relation to them by John Muir. Like the Gariwerd, or Grampian, wetlands, the Stirling Range wetlands have been neglected or misrepresented from the first Englishmen to note them, through a geologist reconnoitering them, to recent writers and photographers representing them in word and image.

As with the Grampians wetlands (as we saw in the previous chapter), the present chapter aims to set the record straight by rereading the record critically from a wetland-friendly point of view within the transdisciplinary environmental humanities. Following on from previous discussions of the uncanny, both in previous chapters of the present volume and previous volumes (Giblett 1996, 2019b), this critical rereading includes focusing on the uncanny aspects of wetlands, including their womb-like spaces and roles as creators of new life. Along these lines, the present chapter celebrates the Stirling Range wetlands in the recent environmental artwork of local resident,

Joanne Francis, who lives next to Woogenellup Lake south of Stirling Range, including her uncanny depiction of a wetland in a painting and her womb-like baskets.

After relating early European exploration, discussing critically landscape aesthetics and introducing a geologist's reconnaissance of the Stirling Range, the present chapter showcases the stunning botanical biodiversity of the Stirling Range in what could be called "Western Australia Felix," Western Australia the blessed, to place beside its eastern Australian counterpart in Gariwerd, or the Grampians, and western Victoria. Yet whereas Major Mitchell called western Victoria "Australia Felix," Australia the blessed, because of its pleasing prospects (in both the spatial and temporal senses) for pastoralism, "Western Australia Felix" is what the introduction to the official natural history of the Stirling Range calls "an amazing treasure house of biological riches" with more than 1,500 species of wildflowers, more "than in the entire British Isles," 87 of which occur nowhere else (Underwood and Burbidge 1993, 3). Yet rather than a treasure house built by human hands for exhibiting and viewing desiccated botanical specimens in the display cabinets of a herbarium or museum, or in the species lists and illustrations of the dead leaves of a book, "Western Australia Felix" is a biologically rich habitat for living native plants and animals created by ancestral beings (Leighton 1993, 31–32). At the end of 2019 the Stirling Range National Park was devastated by a bushfire sparked by a lightning strike on Boxing Day. More than 40,000 hectares were blackened. Some species may not survive, or if they do, recover (Logan and Dobson 2020).

EUROPEAN EXPLORATION

The first Englishman to record the existence of the Stirling mountain range was probably Captain Matthew Flinders, the circumnavigator of Australia and its coastal cartographer. As he was sailing in his ship *Investigator* off the south coast of Australia in 1802 he recorded sighting "a chain of rugged mountains" (cited by Hall, Leighton, and Willis 1993, 35). Continuing in the same vein, Flinders named "the eastern and highest peak" "Rugged Mountain," now known as "Bluff Knoll." By contrast, nearly three decades later the Swan River Colony Regional Representative in the region, Alexander Collie, recorded the Aboriginal name for the range in his exploration journal of 1831 and the Aboriginal names for some of the main peaks, including Toolbrunup that survives to this day (Leighton 1993, 27; Hall, Leighton and Wills 1993, 35).

Not continuing in the same vein as Flinders and Collie, the Swan River Colony surveyor general, John Septimus Roe, renamed what he described as "an elevated picturesque range of mountains" as "Stirling Range" (Roe, 2014, 99). Roe related later in his journal that:

> these remarkable & picturesque mountains being as yet unknown collectively by any distinguishing appellation, & His Excly having kindly consented to my conferring on them a name I called them the "Stirling Range." (Roe 2014, 100)

A footnote here supplied by the editor of Roe's journal points out that "in Roe's fieldwork notes [. . .] upon which he based this account, there is no mention of naming the range in honour of Stirling" (Roe 2014, 100, n.162). In writing up his fieldwork notes into his formal journal he *re*named the range as it was known collectively by the distinguishing appellation of its Aboriginal name of "Koi Kyeunu-ruff" as recorded by Collie four years earlier (Leighton 1993, 27).

Roe goes on to relate that "several of the most conspicuous elevations on the range, whose native names were not very poetical also underwent a change" at his forced instigation, but not Toolbrunup that Roe refers to twice by name (Roe 2014, 100). It is debatable that the "native names were not very poetical" as the Aboriginal name for the range translated roughly means "place of ever moving about fog and mist" (Olver and Olver 1998, 3; see Leighton 1993, 27). Not only is this "an apt name indeed for the Stirlings," as the Olvers go on to point out, but it is also a far more poetical name than the name Roe chose after his visit there in 1835 in honor of his boss and travelling companion, his Excellency the governor of the Swan River Colony, Captain Sir James Stirling, who kindly consented and mock-modestly went along with the name change.

The name change is enshrined cartographically in Arrowsmith's map of 1848–1949 based on Roe's surveys of 1835 (Arrowsmith 1848–49) Arrowsmith's previous map of 1833 followed in Flinders' footsteps and referred to the Stirling Range simply as "Rugged Mountains" (Arrowsmith 1833). Both maps show an unnamed Woogenellup Lake south of the Stirling Range in the middle of (and in the caption for) "sandy plains clear of trees."

ABORIGINAL SIGNIFICANCE

Like Roe, both maps also retain the apt and poetic Aboriginal name of "Toolbrunup" for one of the mountains that means "drizzle carrier" (Leighton 1993, 27). This name makes the mountain sound like a character in the Old English poem *Beowulf.* Indeed, Toolbrunup is an actor, if not a central character, in the very old story of the life of the "place of ever moving about fog and mist." *Beowulf* is for J. R. R. Tolkien (2006, 30–31) "the most successful Old English poem" and "a heroic-elegiac poem" that tells in part the story of Beowulf, a marsh monster and his mother. The name of Toolbrunup tells the very much older heroic-elegiac story of the geology of a mountain, a mountain range, and its motherly marshes (considered as monstrous by some). The core of the Stirling Range is thought to have been formed about 1,150 million years ago (Semeniuk 1993, 15). Toolbrunup on this view is not the scenic and static backdrop for human drama as in white history, nor a singular, sublime, and statuesque figure as in white geology, nature idolatry, and secular theology (as we will later in the present chapter).

A strong case could be made for re-renaming the range back to its original Aboriginal name, especially as Stirling was the leader of the armed party including Roe that committed the Pinjarra massacre about a year earlier in October 1834 (Owen 2019) and with the precedent recently set in Western Australia with the re-renaming of "the King Leopold Range," named after a Belgian monarch, racist, and slaver, back to its original Aboriginal name (Australian Associated Press, 2020). A precedent was also set in Victoria with re-renaming the Grampians as Gariwerd (as we saw in the previous chapter). Western Australia should follow suit with its eastern Australian compatriot and peer. Names and naming are powerful; renaming is empowering.

The Aboriginal names for Toolbrunup and the range foreground the importance of water for life on the driest, continuously inhabited continent on earth, and for the biodiversity of the plant communities and the habitat they provide for animals in the area. Two eminent botanists, Greg Keighery and John Beard (1993, 43 and 49), relate that:

> rainfall is raised by clouds encountering the high peaks, enabling woodlands, swamps and thicket to grow where one would expect mallee shrublands and heath [. . .] The [national] park lies at the edge of the wheatbelt, but is close enough to the coast to allow the mountains to moderate the climate. (Keighery and Beard 1993, 43 and 49)

A similar situation occurs with the topography and location of Gariwerd on the edge of the dry Mallee and Wimmera in western Victoria (as we saw in the previous chapter).

The name change of the range from the poetic Aboriginal name, through the prosaic "Rugged Mountains" to the Stirling Range, is enshrined ultimately in "the Stirling Range National Park proclaimed in 1913, only the third national park in Western Australia" (Hall, Leighton and Wills 1993, 35–36, 39; see also Olver and Olver 1998, 20). Like Gariwerd, it is a mountainous national park in keeping with the American model devised and propagandised by John Muir, "the patron saint of national parks" (Giblett 2020d, chapter 5), as a locus of the sublime (for further discussion of the Aboriginal significance of the Stirling Range, see Leighton 1993, 27–34).

LANDSCAPE AESTHETICS

Also in keeping with Muir's American model, and with Gariwerd, the aesthetics of the mountains were paramount in proclaiming national parks in the United States and Australia, and continue to be important in current perceptions and valuations of natural scenery, including of the Stirling Range National Park. The official natural history of the Stirling Range, *Mountains of Mystery*, waxes lyrical about the visitor to the Park experiencing "the mystery of the mountains" on "a misty mountain top" (Underwood and Burbidge 1993, 3). Mountains of mist, though, are a more apt sub-title for the Stirling Range than the ersatz religiosity and secular theology of the sublime evoked in the mysterious mountains following in Muir's mystical, mountain-climbing footsteps (see Giblett 2020d, chapter 5).

Underwood's and Burbidge's reference to mist does retain a trace of the Aboriginal name for the range though. It acknowledges the importance of water for the life and biodiversity of the place (especially its wetlands). Yet they can't resist airing their aesthetic prejudice in favor of the mysterious misty mountains by having a sly dig at "the featureless bush," despite the biodiversity of "the bush" (Underwood and Burbidge 1993, 3). They are following in the footsteps of an early twentieth-century geologist who had equally disparaging things to say about the "sombre dwarfed vegetation" of "the featureless bush" (as we will see later in the present chapter).

Muir was not alone in the nineteenth century in valorizing the mountainous as evocative of the sublime. Indeed, it was a commonplace of nineteenth-century landscape aesthetics, such as for the art historian and cultural tourist, John Ruskin, who goes one step further than Muir. For Ruskin, "mountains are the beginning and end of all natural scenery" (cited by Olver and Olver 1998, xii). On this view, wetlands are before the beginning and after the end of "natural scenery." More precisely, wetlands were in the beginning of creation as God's first and best work (*Genesis* 1: 2; chapter 2). They may be in the end (of creation) too. For Ruskin, though, they are outside the history and range of natural scenery. They have no proper place in conventional European landscape aesthetics and have not had one in mainstream western culture more generally for over a millennium. By implication, wetlands are the beginning and end of all improper natural obscenery. Ruskin was following in the footsteps of his compatriot Charles Cotton of the previous century. For Cotton in his poem, *The Wonders of the Peak*, "black heaths" are "nature's shames and ills" (cited by Giblett 2011, 65). The black waters of heaths and other wetlands are obscene and so should remain hidden, out of sight and out of mind.

Rob and Stuart Olver's coffee table book cum tourist guide to the Stirling Range and neighboring Porongorups lavishly illustrated with color photographs cites Ruskin as the epigraph to their book. Following in Ruskin's footsteps, the Olvers largely ignore the natural obscenery of wetlands in the Stirling Range National Park as they are fixated on "the beginning and end of all natural scenery" in the mountains. On their detailed map of the Stirling Range they show and name several swamps and lakes, such as Pillenorup Swamp, as well as showing an unnamed Woogenellup Lake outside the National Park (Olver and Olver 1998, 8–9). They later state that "freshwater sedge swamps" are one of a variety of "minor plant communities in the national park" that contribute "greatly to the plant diversity on show in the Stirlings" (Olver and Olver 1998, 52). This ranking is misleading and flies in the face of the botanical references in several chapters of the official natural history of the Stirling Range that the Olvers cite occasionally when it suits them and when it conforms to, and confirms their pro-mountain, anti-wetland prejudices.

Writing on plant communities, Greg Keighery and John Beard (1993, 43 and 47), two of Western Australia's most eminent botanists over the past sixty years, regard freshwater sedge swamps, such as Pillenorup Swamp, "the only large unwooded, freshwater swamp in the park," as one of "five major plant communities" of the Stirling Range, and not as a minor plant community as the Olvers do. Pillenorup Swamp, other freshwater wetlands and "salt lake communities" are shown on Keighery's and Beard's map of "Stirling Range National Park Vegetation" (Keighery and Beard 1993, 46–47). Pillenorup Swamp is also shown on both of Arrowsmith's maps (Arrowsmith 1833, 1848–49). By organizing its discussion of the flora of the Park based on the major genuses of "Eucalypts," "Orchids," and so on according to scientific speciality, and thus abstracting them from their ecosystems such as wetlands, *Mountains of Mist* makes it difficult to appreciate the contribution freshwater sedge swamps and salt lakes make to the botanical biodiversity of the Park.

Despite this contribution and showing Pillenorup Swamp on their map, the Olvers do not acknowledge in writing the botanical status of freshwater sedge swamps as a major plant community, nor depict their botanical richness by showing them in a photograph. What do they look like? Who knows? They are obscene, out of sight, out of mind, not on show for the viewer of the photographs in their book. Apart from Quaderwardup Lake, none of the water bodies shown on the Olvers' detailed map is considered worthy of photographic depiction. Quaderwardup Lake is only included on three occasions because in the first instance it forms the mid-ground of "a moody view [. . .] in the Stirlings" of one photograph with the lake and mountains depicted and captioned in the melancholic register at twilight (Olver and Olver 1998, 22–23). It is included in a second photograph with "the serrated edge of the eastern Stirlings" in the background (Olver and Olver 1998, 102–103).

Quaderwardup Lake is included in a third photograph as it forms the foreground of a double-page photograph with Toolbrunup in the background "in a scene reminiscent of the European Alps" (Olver and Olver 1998, 105–107). The Australian landscape is subsumed to the conventions of European landscape aesthetics and its pinnacle in the Alps with mountains confirmed as "the beginning and end of all natural scenery," albeit with an alpine-looking lake (but not a wetland) thrown in for good measure.[1] The surface of the body of the earth is spread out for display and visual pleasure, and its depths denied. Ruskin would be proud of his progeny singing word-for-word and in tune from his song-sheet and imaging photo-by-photo from the templates in his aesthetic picture pattern book of mountains.

The lake in this double-page photo is also aestheticized in a scene reminiscent of the centrefold in a pornographic magazine. The Australian landscape is subsumed to the template of the European aesthetic of the nude portrait and its pinnacle or pits in pornography in which the surface of the body of sexual desire is spread out for display and visual pleasure in the dead matter of a photograph, and so its living depths are denied. By aestheticizing a lake and not photographing wetlands, Rob Olver's landscape pornography reproduces Ruskin's famous horror of female genitalia and pubic hair as wetlands are wombs of new life (as we will see later in the present chapter). Ruskin would be proud of his progeny for both reproducing his privileging of mountains as the beginning and end of natural scenery, and re-presenting his horror of female genitalia and the womb as the beginning and end of natural obscenery.

Unlike "Australia Felix" with its colonial landscape painter of Eugene von Guérard (as we saw in the previous chapter), "Western Australia Felix" did not have an equivalent colonial landscape painter in the nineteenth century. The two mountainous areas on opposite sides of the continent show some similarities now as tourist sites and national parks with wetlands, but the Stirling Range has no counterpart to Gariwerd's von Guérard (and vice versa). If a colonial landscape painter had worked in the area, he would have probably depicted the Stirling Range in similar ways to von Guérard's depiction of Gariwerd with sublime mountains dominating the painting, possibly with slimy swamps in the foreground or mid-ground.

Marianne North, the noted botanical and occasional landscape artist, traveled in the area in the 1890s. She described it, and some of its flora, in her autobiography (as we will see shortly), but she does not seem to have depicted the area in landscape painting. Unlike von Guérard with his grand, sweeping landscape paintings of vast mountainous panoramas, North's landscape paintings take a more close-up view of the land with trees in the foreground and mountains relegated to the background. What a pity that North did not paint any landscapes of the area she traveled through. Her work is housed in her own gallery at Kew Royal Botanic Gardens in London (North, n.d.).

GEOLOGY

The Olvers do not follow in von Guérard's footsteps in depicting the wetlands of the Stirling Range. In following Ruskin by not doing so, they also follow W. G. Woolnough, "lately professor of Geology, University of Western Australia," in his "geological reconnaissance" of the Stirling Range published in 1920, seven years after the Stirling Range National Park was proclaimed. Like the Olvers, Woolnough also follows Ruskin as wetlands remain hidden in his account. Geology is not immune from the contagion of the conventions of the European landscape aesthetic. It is also blind to botanical biodiversity, despite already having been documented by the government botanist (as we will see shortly).

The Stirling mountain range for Woolnough "forms a unique feature in the geology of Western Australia." For him this is partly so, because, as he goes on to relate:

> the mountains rise abruptly from level plains, mostly "sand plains," which, with an average altitude of about 900 feet above sea-level, stretch far and wide in almost every direction. To the north, as far as the eye can reach, even from the highest summits, these plains continue without interruption. (Woolnough 1920, 80)

Woolnough does not so much see with the eye as reach out with it to try to grasp the immensity of the plains. The eye becomes an instrument of touch and possession. It is not an organ for the reception or appreciation of a scene or a view, let alone of a plant or a flower. The eye is the dominant sensory organ, not for seeing, but for apprehending and grasping the immensity of the plains in an extension of, or prosthesis for, the hand in the drive for technological mastery over the land.

The plains continue without interruption, but not without variation though, as Woolnough (1920, 82) goes on to describe how:

> to the north of the Stirling Range, the plain above mentioned is simply crowded with salt lakes of all shapes and sizes. From the high peaks of the range they can be counted literally by the score, their white and glistening surfaces being in marked contrast to the sombre dwarfed vegetation of the sand plain.

"Sombre dwarfed vegetation" is a bit disparaging and only a slightly better attempt at describing heathland and banksia woodland than the derogatory term "scrub." It is on a par with Underwood's and Burbidge's "featureless bush." "Sombre dwarfed vegetation" also evokes the grotesque as if the vegetation is made up of monstrous and mythical beings.

This low vegetation in Woolnough's eyes is endemic to the south-west botanical province of Western Australia, an internationally recognised biodiversity hotspot and the only one in Australia out of twenty-five around the world. The introduction to the official natural history of the Stirling Range states that it has:

> long been recognised as an amazing treasure house of biological riches [. . .] The number and beauty of the wildflowers are staggering—more species occur in the Stirling Range than in the entire British Isles—and 87 plant species found in the Stirling Range occur nowhere else on Earth. (Underwood and Burbidge 1993, 3)

Over 1,500 species of plants in total have been recorded in the Stirling Range, including "an astonishing 123 orchid species" (Olver and Olver 1998, 50 and 57–59; Keighery and Beard 1993, 43, 48–49; Brown 1993, 61). Rob Olver concludes that "the Stirling Range National Park is a wildflower and plant reserve of international significance" (Olver and Olver 1998, 59). As well as plant species, endemic animal species are also to be found in this area (Olver and Olver 1998, 35, 39–49). It is indeed "Western Australia Felix" as a living, biologically rich habitat.

Woolnough cannot see or acknowledge this botanical biodiversity, nor the wetlands around him, but takes in the distant view from the mountain range of salt lakes, sand plains, and stunted vegetation of "Western Australia Execratus," Western Australia the cursed. This botanical richness was established by the government botanist from the 1840s, seventy-five years before Woolnough's paper was published. No doubt Aboriginal people were aware and appreciative of it for tens of thousands of years before that. The geologist Woolnough did not read the work of the Government Botanist, James Drummond, perhaps another sad and sorry case of siloed disciplines, even occurring in the same state of Western Australia, studying the same object, but without a historical or transdisciplinary perspective. In 1843 Drummond "was the first [white person] to realize the huge plant diversity harboured in the Stirlings," as Rob Olver puts it (Olver and Olver 1998, 16, 49–57). In 1920 Woolnough was not the last white person to fail to appreciate the huge plant diversity harbored in the relatively safe confines of the Stirling Range National Park. In the 1990s, for instance, it took some time for Ailyeen Sands to do so, changing and culminating in her ecstatic encounter with the orchids of the area (Ryan 2012, 188–189 and 205–206).

Fifty years after Drummond, the noted artist Marianne North (1893, 149–150) stayed in Albany and then traveled overland to Perth. She describes banksias as "quite marvellous," hakeas as "one of the remarkable plants in the world" and "large tracts of sand" as "wearisome," like Woolnough, "except that it gave one time to see the endless variety of flowers," unlike Woolnough who could only see monochrome, mini-monstrous, and undifferentiated "vegetation," another sad and sorry case of not being able to see the trees for the wood, the individual plants for the overall vegetation.

The drab green of the "sombre dwarfed vegetation" contrasts starkly for Woolnough with what he calls the "white and glistening surfaces" of the salt lakes. Woolnough (1920, 82) goes on to contrast the north and south sides of the Stirling Range, and to prefer the latter to the former:

> On the southern side of the range, and between it and the Kalgan River, there stretches a line of salt lakes, quite distinct in many ways from those to the north. The depressions in this system occur singly, and instead of being distributed promiscuously as appears to be the case with their northern brethren, they are arranged in a roughly linear fashion.

It is very considerate of the brotherly, southerly salt lakes to be ordered in a disciplined line, unlike their disorderly and sluttish northern sisters (rather than brethren) that have bred illegitimately and willy-nilly across the sand plains. Woolnough genders Western Australia Execratus' as Major Mitchell did with "Australia Felix" (as we saw in the previous chapter).

Woolnough contrasts the highlands and the lowlands, and values the high over the low, the vertical mountains over the horizontal plains in a stereotypical privileging in accordance with the conventions of the European landscape aesthetic and its spatial hierarchy of land values. While the salt lakes' "white and glistening surfaces" are beautiful, the "solitary grandeur" of a single mountain evokes the sublime.[2] Woolnough feminises virginally the former and masculinises personally the latter when he waxes lyrical about "the towering mass of Toolbrunup" and goes on to describe how:

> this magnificent peak reaches 3341 feet, and, rising as it does in solitary grandeur from the almost level plains at its feet, it is one of the most conspicuous and characteristic features of the entire range. (Woolnough 1920, 84; see also 89–90)

The eye, unable to grasp the profligate immensity of the plains, grasps the singular phallic erection of the magnificent, massive mountain up-thrusting from the surrounding plains. Whereas for Roe the entire mountain range in its horizontal extension was picturesque, for Woolnough the magnificent peak of a single mountain in its vertical erection was sublime. Both are singing from the song-sheet of the European landscape aesthetic of the picturesque and the sublime.

The "solitary grandeur" of Toolbrunup is a statuesque figure at whose feet the plains spread out, accept the run-off from the mountain and create lakes. Woolnough goes on to relate that "the country to the north of the Stirling Range forms an area of internal drainage. Its characteristics and the extraordinary abundance of salt lakes have been described above" (Woolnough 1920, 108). As with Gariwerd, and not surprisingly, "the Stirling Range is an obvious watershed for the drainage in this region [. . .] The [fringing] wetlands (or salt lakes) [. . .] are fed by creeks and subterranean discharge," as Semeniuk (1993, 15 and 22) puts it in his more recent study of the geology of the Stirling Range. Nurtured by surface and subsurface living waters these wetlands are uncanny, rather than obscene, even though they are partially hidden.

Woolnough appreciates the glistening white surface of the salt lakes, but overall in his estimation they are failures when it comes to performing their proper function of being rivers that drain the mountains. He goes on to relate how:

> the lakes represent the expiring efforts of a number of small creeks, heading in the Stirling Range to extend across the sand plain and reach the Kalgan or the Pallinup [Rivers]. Originally they may have been successful but as the climate has become progressively drier, their waters have failed to extend beyond the limits of the foothill. (Woolnough 1920, 110)

Woolnough blames the lakes for failing to be rivers that drain the mountains properly. The expired creeks in the salt lakes of the supine plains have failed to satisfy and drain the phallic mountain. They are poor sexual partners for the virile mountain, unlike his prosthetic eye that grasps the phallic mountain and gives it sublimated satisfaction. Woolnough develops an engineering technology of his prosthetic eye grasping the phallic mountain and gives birth out of the simulated womb of his brain box to the sublimated satisfaction of the sublime independent of the uncanny fertility of the earth, especially womby wetlands.[3]

Aboriginal people feminised the range as it has, when viewed from the south, "the outline of the female body [with Bluff Knoll as her head and Toolbrunup as her breasts]. This is known as the Sleeping Beauty" (Leighton 1993, 32; illustrated on 31). To whom the range is known as this female figure in a European fairy tale is not clear, nor is it clear who the handsome prince who would awake the sleeping princess with a kiss. Rather than a passive princess in a fairy tale, she is the active, polymorphously perverse and polysemic feminine figure of the creative and destructive Great Goddess or Mother of the earth story with her wetlands and drylands, highlands, and lowlands. She contrasts with the "towering mass" and "solitary grandeur" of Woolnough's singular sublime and masculinist mountain at whose feet the supine and passive feminised plains with their glistening salt lakes offer inadequate sexual services for, and poor drainage of, the phallic mountain.

The lakes look attractive on the surface in their glistening virginal white, but underneath they do not perform adequate sexual services. They are exhausted and have become old and dried up. Woolnough splits the lakes between the good, young, beautiful white virgin on the surface and the bad, dried up, uncanny old dame in the depths beneath the surface as a defense against the horrors of what he sees as "Western Australia Execratus." He genders and gerontises geology and sexualises the earth by feminising plains and salt lakes and by masculinising mountains and geologists. He does not acknowledge that the drying climate is anthropogenic, partly caused by clearing native vegetation, which was starting to be understood in his day, even though a heating climate caused by burning fossils may not have been.

ENVIRONMENTAL ARTWORK

Contrast Woolnough's masculinist view of the mountains, the lakes, and the plains and his disparaging view of the vegetation in his "geological reconnaissance" with Joanne Francis's portrayal of wetlands in her environmental artwork, such as her painting, "Tranquillity"(figure 5.1). The lower third of the painting portraying the sky reflected in the still water of the wetland dominates the painting. The artist inverts, if not subverts, "the rule of thirds" in conventional landscape painting and photography that devotes the lower third to the darker land and the upper third to the lighter sky. In this painting, the lower two-thirds depicts the lighter sky reflected in the water (albeit with looming dark clouds reflected in the bottom foreground, perhaps presaging some impending doom), while the darker land is relegated to the upper third and the sky is not depicted as such at all. The normative hierarchy in which land is valued over sky and water is subverted. In this painting, the wetland dominates, very unusual in Australian landscape painting of wetlands (see Giblett 2020a). In fact, this is not a landscape painting at all, nor even a wet landscape painting, but a portrait of a wetland as a living being or body with depths in its mirrored surface. It contrasts with Rob Olver's pornographic centrefold of the dead surface of his photograph of a lake.

The bare trunks and limbs sticking out of a clump of grasses and reeds at odd angles in Francis' painting are elongated grotesquely and reflected darkly in the still surface of the water. The spidery trunks and limbs are

Figure 5.1 "Tranquillity." Source: Courtesy of Joanne Francis

either a marshy mother/monster into whose maternal embrace the viewer may wish to enter and be hugged, or the femme fatale of a seductive but dangerous spider woman from whose ravening maw the viewer may want to flee in horror and not be devoured. The wetland is a womb and tomb from which the viewer is reborn, or a tomb in which the viewer might drown and die, or an orally sadistic monster who might consume the viewer (Giblett 1996).

Francis' painting seems to echo and play intertextually with Sidney Nolan's "Carcase in Swamp," now housed in the Tate Gallery in London (Nolan 1955). Whereas Nolan shows the carcase of a dead animal stuck in a swamp in the time of a terrible drought in Queensland in 1952, Francis shows a dead tree stuck in the middle of a living swamp during the wet seasonal cycle. Whereas Nolan's carcase is a sacrificial animal in the desert that dies to atone for sins and to seek forgiveness and redemption from various pagan gods or the monotheistic God, Francis' dead tree is an emblematic plant in the swamp from which new life springs, into which old life dies and is reborn. Whereas in Nolan's painting the empty eye-sockets of the whitened skull stare imploringly at the viewer and the blackened legs of the carcase point defiantly to the sky in a futile and unfulfilled plea or prayer for rain, in Francis' painting the black and white limbs of the dead tree in a water-filled wetland reach out to invite and embrace the viewer.

Francis' painting dials or ratchets Nolan's painting anti-clockwise 90 degrees as it were. In the process, it subverts his grim view of death in a swamp, as there is life and death in her swamp. Whereas Nolan aestheticizes the melancholic carcase in the mournful swamp as an emblem of death (the skull, or death's head, is a memento mori, a reminder of death), Francis anti-aestheticizes the wetland in the monstrous uncanny as a maternal allegory of life and death in endless cycle in utero. Whereas Nolan's painting operates in an aestheticized, cerebral, symbolic, monumental and sublimated upper register, Francis' painting plays in the embodied, grotesque, monstrous, uncanny and desublimated lower bodily and earthly strata.

Developing along similar lines to her painting, Francis's environmental artwork includes weaving baskets that are womb-like receptacles, like wetlands. They create figuratively and literally a space and place to replay being in the womb (in utero) (figures 5.2 and 5.3). These two baskets create a warm and inviting interior space into which the viewer might visually enter and be enclosed warmly in its womb-like place, and then be reborn. They contrast with Woolnough's prosthetic eye grasping the mountain. Rather than a vagina dentata, these baskets portray a warm and inviting vaginal entry to and exit from the womb. The latter basket is fringed with furry flowers in a reconstruction of pubic hair, such an object of horror for Ruskin. Here is the ultimate natural obscenery for Ruskin, the other side of the beginning and end of all natural scenery in the mountains that makes it possible.

Figure 5.2 "Vessel Spanning the Isle." *Source:* Courtesy of Joanne Francis

Figure 5.3 "Protea Swirl." *Source*: Courtesy of Joanne Francis

Figure 5.4 "Nest," an Ephemeral Installation Piece. *Source*: Courtesy of Joanne Francis

Francis's portrayal of the womb in her smaller baskets is obvious in her larger basket (figure 5.3). Francis's adult-size basket is a dry land coracle, an amphibious vehicle, for rebirthing its occupant in the fetal position from its womb-like space and leading her, and drawing the viewer, into the watery womb of the promised (wet)land as portrayed in Francis's painting, "Tranquillity" (figure 5.1). This basket alludes to and plays intertextually with the biblical story of the baby Moses who has hidden in a basket among reeds away from the Egyptians who would have killed him as the Pharaoh had ordered that all Hebrew baby boys were to be drowned at birth. His name means, "I drew him out of the water" (*Exodus* 2: 1–10). Moses was the leader anointed to lead God's chosen people of the enslaved Hebrews out of Egypt on their exodus into the promised land, though he did not end up doing so. He only saw it afar from a mountain (*Deuteronomy* 3: 27).

By contrast, Woolnough is a kind of latter-day anti-Moses who stands on a mountain, but does not see the promised land of the botanical biodiversity in the wetlands and drylands around him at his feet in "Western Australia Felix." The promised land is not to be grasped from afar with his eye, but at his feet to be seen, smelled, and touched in a multisensory, embodied experience of plants and place (including wetlands). He looks from the mountain top at the dry and deserted biblical wilderness of the surrounding sandy plains, salt lakes and stunted vegetation, "Western Australia Execratrus," Western Australia the cursed.[4] It is not the Western Australian equivalent of "Australia Felix" with its pleasing prospect of the picturesque pastoral. Woolnough looks backward to where the chosen people have come from out of the deserted wilderness, not forward to where they should be going into the promised land. He misleads the chosen people of white settlers on a retrograde exodus to the secular, singular, statuesque and sublime mountain of idolatry. He does not lead them into the promised land of the botanical biodiversity and wetlands of the mountain range, "Western Australia Felix" (as Drummond, Keighery, and Beard do). Woolnough is a latter-day Aaron, Moses' older brother, who led the chosen people astray into idol worship, in Woolnough's case of the mountain, in Aaron's, of a golden calf (Exodus 32: 1–5).

Nolan's painting of a silvery cattle carcase in a swamp is an ironic take on Aaron's golden calf and an acerbic comment on the pastoral industry in Australia. From Major Mitchell's "Australia Felix" of the pleasing pastoral prospect of the "gentleman's park estate" to Nolan's depiction of "Australia Execratus" with a cattle carcase stuck in a swamp is only a matter of a season or two of bad rainfall. The sacred cow idolised by the Australian pastoral industry and sustained by the promised land of waterholes and swamps can easily and quickly become a drought-stricken desert with a carcase stuck in a drying up, muddy swamp as in Nolan's painting. This swamp is akin to Arthur

Boyd's paintings of "The Waterhole, Central Australia" and "Irrigation Lake, Wimmera" in far western Victoria (Giblett 2020a, figure 3.3, 58; figure 3.4, 59). Both paintings depict drying up water bodies, the other side literally and metaphorically of "Australia Felix," including flourishing Mt William Swamp as depicted by Eugene von Guérard (figure 4.2).

A much more trustworthy Moses than Woolnough who does lead people into the promised land of "Western Australia Felix" and does not view it from afar is the cultural botanist, John Charles Ryan of New England and New Holland, "the patron saint of plants" (Giblett 2020d, v and 33). Ryan relates a wildflower tour from Perth to Albany as part of the programme for delegates at the 2014 Australian Garden History Conference, subsequently written up and published in *Australian Garden History* (Ryan 2014). The article includes excerpts from North's descriptions of the flora around Albany and reproductions of two of her paintings, one botanical, and one landscape. The article culminates by noting that "the alpine wildflowers of the Stirling Range are [one of] a few of the many beguiling features of botanical intrigue [of an overland journey] from Perth to Albany" (Ryan 2014, 9).

Ryan concludes his article by inviting his readers and fellow wildflower tourists and cultural botanists to "reflect during the overland journey" and to still possibly:

> feel the exhaustion of surveyors, the exasperation of homesteaders, the exhilaration of naturalists, and the euphoria of botanical artists who revelled in the unparalleled botanical richness of the region. (Ryan 2014, 9)

In other words, it is "Western Australia Felix." Ryan does not invite his fellow wildflower tourists and cultural botanists to feel the denigration of a geologist who idolized a sublime mountain and reviled the botanical richness of the region around it, both in the wetlands and on the plains, and for whom it was "Western Australia Execratrus."

Ryan, unlike Woolnough, Ruskin and the Olvers, does not lead people astray into idol worship of mountains and denigration of everything else. Mountains for Judeo-Christian theology are, as one theologian puts it, at "the intersection of heaven and earth (*Exodus* 19: 18; 24: 17) [. . . T]he most sacred space was on the highest mountains" (Sandy 2012, 554). They are not, or should not be, objects of worship per se. They are sacred spaces for communion and communication between the divine and humans (as the verses referred to in the biblical book of *Exodus* demonstrate). They should not, therefore, be worshipped themselves (as Ruskin and the Olvers do) as that would be in contravention and violation of the first two of the ten commandments, "you shall have no other gods before me" and "you shall not make for yourself an image in the form of anything in heaven above or on the earth beneath or in the waters below" (*Exodus* 20: 3–4).

When Ruskin idolizes mountains as "the beginning and end of all natural scenery," they are made to usurp the position and role of God in Judeo-Christian theology as both creator and destroyer of earthly time, natural scenery and "obscene" wetlands. When the Olvers idolise images of mountains in the photographs of their lavishly illustrated coffee table book for the endo- and exo-tourist, they are putting their (polytheistic) gods before the (monotheistic) God. By contrast, for conservation counter-theology, the Great Goddess of the natural obscenery of slimy swamps is the beginning, middle, end and rebirth of life. She is both creator and destroyer in life-giving and death-dealing waterlands.[5]

Along similar lines, when Henry David Thoreau (1982, 613), "the patron saint of swamps" (Giblett 1996, ix and 229–239; chapter 8), announced in his essay, "Walking," that "the most dismal swamp" is the holy of holies for him in which he would "recreate" himself, he was not controverting the holiness and sublimated sacred space of sublime mountains (as he did that too in the mountains of Maine). Rather, he was subverting the idolatrous worship of mountains and the placist denigration (literally "blackening") of slimy swamps as dismal places.[6] He was also proclaiming the slimy swamp as a sacred place at the intersection of heaven and earth for communion and communication between the divine and humans, just like the mountain in *Exodus*.[7] Heaven is not only "under our feet as well as over our heads," as Thoreau (1982, 525) wrote in *Walden*, but also around our feet, and even up to our chin in the slimy swamp, as he also wrote in his journal. He could "fancy that it would be a luxury to stand up to one's chin in some retired swamp a whole summer, scenting the wild honeysuckle and bilberry blows, and lulled by the minstrelsy of gnats and mosquitoes!" (Thoreau 1962, *I*, 141). Thoreau, unlike Woolnough, senses the swamp by smelling the flowers, hearing the insects, and by it touching the surface of his body. He engages in bodily dialogue with the swamp by entering it. He does not try to grasp it with his eye. He has an embodied and multisensory experience of the swamp.

Following in the footsteps of Thoreau who fancied himself chin-deep in the swamp, Aldo Leopold, "the patron saint of marshes," went one step further and deeper into a wetland by wishing he were "eye-deep in the marsh" (cited by Giblett 2020d, 102). Both Thoreau and Leopold wrote a word for wetlands and are environmental apostles. Leopold proclaimed his good news of a land ethic and conservation aesthetic for all natural scenery and obscenery (Giblett 1996, ix and 242–245; 2020d, chapter 7).

Wetlands are wombs of new life and tombs of old life out of which new life springs (Giblett 1996, 163, 176). Wetlands are places and spaces of what Hannah Arendt (1996, 51, 132–133, 146–148) called "natality," a term she coined in her doctoral thesis on love in the late 1920s.[8] Arendt placed natality, or being born, at the centre of her social philosophy of life and love. It was a counter to Martin Heidegger's morbid, solipsistic philosophy of mortality and despair facing the fact of one's own death. Her philosophy of being born (and born again) counters his philosophy of being toward death. Death is a part of life, but new life is born in an eternal return and cycle of birth, life, death, birth, and so on. Arendt also placed natality and being born again into the public sphere (*polis*) out of the private sphere of the domestic household (*oikos*) at the centre of her political philosophy. Conservationists, such as Thoreau (and as we will see in chapter 8), place being born again into the ecosphere of the earthly household (*oikos*) in ecology and economy at the centre of their environmental philosophy.[9] Activists, such as Thoreau (as we will also see), place being born again in the public sphere at the centre of their political philosophy. Thoreau the political conservationist did both. Wetland conservationists, such as Thoreau, place being born again in swamps at the centre of their aquaterratological philosophy and practice. Wetland conservation is aquaterranatality. It is depicted in Mules' and Francis' artwork of embryo-shaped wetlands and baskets.

Wetlands are fertile feminine spaces, which is mainly why patriarchy has denigrated (literally "blackened") them as dead black waters figuratively. Wetlands are living black waters that patriarchy has also blackened literally by destroying them. Industrial agriculture and urban development degrade them into the dead black waters of modern wet wastelands (as we will see in chapter 8). They are not a pure, watery womb of life associated with vitality and nourishment—in short, living waters—but a sullied, muddy tomb of death associated with decay and excrement—dead black waters (Giblett 1996, xi–xii, 135). Denigration through destruction to conservation marks the sad and sorry (his)story of wetlands for over a millennium in mainstream western culture. Wetland conservation protects current wetlands and rehabilitates degraded ones. By contrast with industrial agriculture, regenerative agriculture rehabilitates and restores remnant and degraded wetlands on farms. Conserving water in wetlands regenerates the soil. Rehabilitating remnant wetlands or creating artificial ones brings the dying dry land back to life and health. Healthy land is where water is held, conserved and managed sustainably in the long- and short terms during drought or "dry spells." Regeneration through rehabilitation to conservation marks the hopeful story of wetlands on farms in the new millennium.[10]

Wetlands are the first home, or womb, of life on earth, but they also evoke horror in the minds of many males so they are arguably the uncanny place par excellence and they are the home of monsters in the same male minds (Giblett 1996, chapters 2 and 8). Woolnough expresses horror at the uncanny, not only when he makes the vegetation into mini-monsters and feminises the lakes, but also when he splits the lakes between the good, beautiful young white virgin on the surface and the bad, dried up uncanny old dame in the depths beneath the surface. This splitting is a defense against the horror of the uncanny ostensibly by dividing and conquering it. By contrast, Joanne Francis' portrait painting of a wetland and her woven baskets celebrate fascination with uncanny womby wetlands. They invite the viewer to be drawn into her painting and baskets, to bathe in their life-giving and death-dealing water literally or metaphorically, and to be reborn from them.

NOTES

1. For the cultural history of the Alps transformed from "strange, horrid and fearful crags" in the seventeenth century to (being "naturalized" as) the pinnacle of natural scenery in the nineteenth century, see Williams (1973, 128).
2. For further discussion of the sublime and the beautiful drawing on the work of Edmund Burke and Immanuel Kant, see Giblett (1996, chapter 2).
3. In short, "a bachelor machine for a bachelor birth." For further discussion, see Giblett (2008c, chapter 2).
4. For further discussion of the biblical wilderness and desert, see Eggleston (2012, 843–847).
5. See Giblett (1996, chapter 6).
6. For further discussion of placism, see chapter 1 of the present volume and Giblett (2019b, 3 and 34), and of denigration as blackening, see Giblett (2013a, 100–101).
7. For further discussion of slime (critiquing the misogynist work of Jean-Paul Sartre), see Giblett (1996, 39–47).
8. For Arendt's philosophy of love and her relationship with, and views of, Heidegger, see Kristeva (2001, xv, xvii, 8, 44–46, 48, 66–67, 141, 239–240, 243 n27 for Arendt's outrageous and scathing "true story of Heidegger the fox," 246 n75) and Eilenberger (2020, 192, 240, 379 n22, 24, plus photos of Heidegger, the sly fox).
9. For *polis* and *oikos* in both senses drawing on Arendt's work, see Giblett (2011, chapter 2).
10. For "regenerative agriculture" and the stories of wetlands on farms (including his own), see Massy (2020, 63, 130–132, 136–137, 139, 142, 217, 219, 224, 227–228, 312).

Part III

WETLANDS AND CITIES

Chapter 6

The Birth of Sydney and the Death of Its Wetlands

Water is a perennial problem in the driest and longest continually inhabited continent on earth by the oldest surviving culture. Too much water in flood, too little of it in drought, highlight the problem of water, and differing quantities of it, in the right or wrong places, at the right or wrong time. The spatial and temporal variability of water are prominent with wetlands as they are in-between land and water, and even mix earth and water in mud, morasses, marshes, swamps, slime, and so on, with greater or lesser earth and water, in greater or lesser amount and extent over time and in space. Wetlands have had a precarious place in Australia since settlement as many studies show, as well as in the history of water in Australia and several of its cities as many studies also show.

A recent history of water in Australia, for instance, acknowledges that for the first colonial continental settlement of Sydney that eventually grew into Australia's biggest city the nearby wetlands were an important source of water, but were quickly degraded, destroyed and largely forgotten (Cathcart 2010). The same situation applies with the settlements and cities of Perth and Melbourne with nearby wetlands initially providing water for humans and stock, plus feed for the latter, and then they were degraded and destroyed, including being aestheticized and landscaped into parks and gardens. Recent wetland cultural studies in the transdisciplinary environmental humanities on the birth of Perth (Brady and Murray, 2020; Ryan, Brady, and Kueh, 2020) and Melbourne (Giblett, 2020c) acknowledge the wetlands that were present on settlement in these two cities and that are now absent, mourn the loss of these wetlands, call for the conservation of those that remains, propose memorials for those that have been lost, and interpretation of the history and significance of all of them for the residents of both cities.

Melbourne's lost wetlands are largely not commemorated and remembered in public installations or artworks. Perth's lost wetlands are commemorated in memorials and were remembered in an exhibition mounted at Perth Town Hall in September 2014. The materials from this exhibition are now available online on the Western Australian Museum website (Brady and Murray, 2020; Ryan, Brady, and Kueh, 2020). Until the recent exhibition by the Royal Historical Society of Victoria in early 2020 devoted to one lost wetland in Melbourne (discussed in chapter 7), Melbourne's lost wetlands have not been commemorated in exhibitions.

To the best of my knowledge, a similar situation applies to Sydney's lost wetlands with no exhibitions or memorials devoted to them to date. A plaque in the pavement and a permanent exhibition in Martin Place commemorates the Tank Stream (Hinkson 2001, 2–4), but nothing in Hyde Park commemorates the swamp that fed the stream. The present chapter on Sydney and its wetlands proposes that the same courtesy extended to Perth's wetlands be extended to all the largely forgotten wetlands of Sydney. The birth of Sydney, Melbourne, and Perth near wetlands meant the death of the wetlands that were close to the early colonial settlement. For the colony to be established, wetlands were ruined and destroyed. Remnants of wetlands can be found further afield on the outskirts of all three cities and in inland areas of all three states of which they are the capital city (as we have seen in the two previous chapters in the cases of Victoria and Western Australia).

In the case of New South Wales the state government recently purchased 153,415 hectares of private land that was a pastoralist station for a new national park in the state's far north-west. It is the largest single land purchase of private land for conservation in the state's history. It stretches across outback channel country, comprises parts of the floodplain of the Bulloo river, and includes "ephemeral wetlands and landscapes" listed as nationally significant that were not included previously anywhere else across the state's national parks (Readfearn 2020). Wetlands were not included in the state's national parks either because they were not considered worthy of such status (only mountainous or bush areas were for a long time), or because they had been destroyed (as we have also seen in the two previous chapters in the cases of Victoria and Western Australia).

TANK STREAM

In Sydney, the Cadigal, or Tank, Stream that "used to emerge from swampland at present day Hyde Park" (Falconer 2010, 33) "provided an important water source for the Cadigal [or 'Gadigal'] people, the traditional owners of Sydney Cove" (Hinkson 2001, 3; for discussion of the "Gadigal," see

Attenbrow 2010, 24–25). The colonial settlement that grew into the city of Sydney drew its initial water supply from this Stream. When Governor Arthur Phillip chose the site for a colony in 1788 the proximity of the Tank Stream was one of the determining factors for choosing it over Botany Bay that was the initially designated site and where Captain James Cook had landed in 1770 (Korporaal 1991, 28; Hinkson 2001, 3).

Another determining factor was Sydney harbor. Paul Carter (1987, xiv) describes how "Governor Phillip founded the first settlement (the future Sydney) at Port Jackson, a 'capacious' harbor immediately to the north of Botany Bay." Port Jackson supplied what Botany Bay lacked on two counts: a capacious harbor and a water supply. Phillip stated that "I fixed on the one [site] that had the best spring of water and in which ships can anchor [. . .] close to shore" (cited by Statham 1989, 9). For Statham "a safe accessible anchorage would have been his prime consideration" as he was a naval captain. Statham is countering other historians who emphasize the water supply as the prime consideration. Martin (1989, 48; cited by Statham 1989, 9) argues that Phillip abandoned Botany Bay for Port Jackson because he "found its low ground did not provide enough drinking water for the number under his command," over one thousand. Ironically, this "low ground" later became a water source for the settlement. Along similar lines to Martin, Fletcher (1989, 52; cited by Statham 1989, 9) relates that "one of the coves [in Sydney Harbour, that is, Sydney Cove] possessed what Botany Bay lacked, namely a stream for drinking purposes, and here colonization began" with the founding of the colony of New South Wales. Or perhaps more precisely, the colony began with the colonization of the stream and the harbor, and of Australian "nature" more generally. Planting a flag and reading a proclamation were ceremonial and secondary after the primary colonization of the place.

As Lucy Hughes Turnbull suggests (1999, 58), the "deep water" harbor "suitable for shipping" in Sydney Cove, as well as the fact that there were "no mangroves" here unlike Botany Bay and that there was "plentiful fresh water" in the Tank Stream, were all important considerations for Phillip choosing the site. Not surprisingly, the Tank Stream was perhaps the first site of tension or conflict between Aboriginal people and the British invaders. Turnbull (1999, 31) relates how, "within three days of the First Fleet's arrival at Sydney Cove, the Aborigines showed displeasure at the clearing of the land around the freshwater stream that would become known as the Tank Stream as the area had, up until then, been one of the most bountiful sources of game and fresh water in the Sydney area." Whereas Botany Bay had what Cathcart (2010, 11 and 23) says Phillip would regard as "malignant marshes," and what we would today regard as "magnificent wetlands," Sydney had potable water flowing from benign swampland into the Tank Stream. What Botany Bay was lacking, the Tank Stream supplied. Without this stream, the colony would not have taken place when and where it did. The colony took the place of the traditional owners from them; decolonization would give the place back to them.

In an understandable and laudable move to try to protect "Sydney's precious water supply," as Korporaal (1991, 29) calls it, Phillip "enacted Australia's first environmental laws, declaring a 'green belt' 15 m wide on either side of the steam" with severe penalties for polluting it (Hinkson 2001, 4). Yet this "green belt" hardly embraced what Karskens (2009, 74) calls "the 200-foot wide [61 meter-wide] corridor of trees which shrouded the Tank Stream." The Tank Stream was ghostly, perhaps a premonition of the swampland from which it emerged becoming eventually a ghost swamp. Rather than "green bans" being an innovation of the 1970s instituted by Norm Gallagher and the Builders Labourers' Federation to protect the old sandstone buildings of the Rocks and other areas in Sydney, they were participating in a nearly century-old movement begun by Phillip. After Phillip left Australia in 1792, the stream was polluted and "became an open sewer" as Korporaal (1991, 29) puts it, the sad and sorry state and fate of many water sources and wetlands near urban developments both at "home" in the colonial centers of power and "abroad" in the colonies (such as Melbourne; see Giblett 2020c).

In about 1848 an artist gave an impression of the Tank Stream in which they depicted it nostalgically and bucolically using the conventions of the picturesque in the European landscape aesthetics, overlooking (in two senses) the polluted and undrinkable state of the water, and painting over the history of the Stream (State Library of South Australia n.d.).

In about 1842 the artist John Skinner Prout depicted the same view in water-color (State Library of New South Wales n.d.b). In about 1852 the artist John Black Henderson depicted again the same view in water-color (State Library of New South Wales n.d.a). In about 1874 Thomas Heawood engraved Prout's painting for the second volume of *Australia* by Edwin Carton Booth (National Library of Australia, n.d.; reproduced in Korporaal 1991, 29). This view had become a cliché. It was reproduced in a later post-card supposedly showing "The Old Tank Stream" in 1819 (National Museum of Australia, n.d.), look after it had ceased to be a stream of living water. The pictorial depiction of the Tank Stream became a serial nostalgic wet day dream of the time in which it was not polluted and of the place as it might have been (and might still be today).

FROM SWAMP TO PARK

After the Tank Stream was polluted, settlers had to look further afield for a source of fresh water, first to the nearby Lachlan Swamp (present day

Centennial Park) that water carters used as a source of fresh water to supply the settlement, or at least the wealthy members of it (Cathcart 2010, 37). Cathcart (39) relates that "the swamps were a system of ponds and marshes three kilometres long, with a catchment eight times larger than the valley of the Tank Stream. The main pond was just three kilometres south of Sydney Cove, in the area now occupied by the undulating lawns of Centennial Park." Yet Centennial Park and its landscape designers and gardeners invaded land that was already occupied by Aboriginal people, landscaped the swamps into undulating lawns and "ornamental lakes," and so re-occupied the area. The idea of Lachlan Swamp now being "occupied by undulating lawns" smacks of the idea that the British occupied an unoccupied Australia, terra nullius, when in fact they invaded the occupied homelands of Aboriginal peoples, just as they invaded their occupied wetlands, such as the Lachlan and Botany Bay Swamps. The park landscape aesthetic beloved of landscape architecture and gardening was a tool, instrument and weapon of invasion and colonization.

In their history of Centennial Park, Paul Ashton and his colleagues at the University of Technology Sydney describe how "the Lachlan Swamp [was in fact] an extensive swamp, 3.6 kilometres in length with a catchment area of between 486 and 600 hectares. [It] was the closest in this chain of swamps to Sydney" (Ashton et al. 2013, 13). Cathcart (2010, 39) relates that "this swamp itself was part of a much larger chain of lagoons, marshes, heathlands and sand dunes that extended all the way to Botany Bay. Ironically, these wetlands were the far side of the same 'unhealthy' swamps that had so dismayed Arthur Phillip." Similarly, the Moonee Ponds Creek in Melbourne was a large chain of swamps, lagoons, marshes, heathlands and sand dunes. It is now a drain, though moves are afoot to restore it partially to its former wetland glory in wetlandscape architecture (as we will see in the following chapter).

Water from the Lachlan Swamp was stolen away to Sydney, initially by carters, later by tunnels and pipes (Cathcart 2010, 46). John Busby, a mineral surveyor and civil engineer, described the water of Lachlan Swamp as "perfectly transparent and colorless, free from every taste and smell, and so soft to be fit for washing and every other domestic purpose" (cited by Ashton et al. 2013, 17). This description is hardly surprising as the water of wetlands is typically filtered by vegetation and, in the case of Sydney's Lachlan Swamp, by sandstone that "acts as a giant filter," as Falconer (2010, 35) puts it.

Busby proposed the construction of a tunnel, commonly called "Busby's Bore," from the Lachlan Swamp to Sydney. It took ten years to construct and was completed in 1837 after a decade-long tortuous saga involving the use of recalcitrant convict labor and because of Busby's poor supervision as he hated going underground (Cathcart 2010, 41–43; Ashton et al. 2013, 16–17 and photos). From 1837 to 1859 Lachlan Swamp was the main source of

Sydney's freshwater supply as Turnbull (1999, 392) relates. Busby's Bore was the means to convey it, and not the source itself, as Turnbull (1999, 263) earlier suggests—the swamp was the source. Yet, despite the importance of the swamp as the main source for supplying fresh water to the city, by the late 1840s, the Lachlan Swamp was drying up. By the 1860s it had ceased to be wet land and had become "a tract of barren sandhills" in the words of an eyewitness (cited by Cathcart 2010, 45). From fertile watery womb to dry barren womb, or what Ashton (2008) calls "sandy, swampy wasteland," marked the sad and sorry fate of the Lachlan Swamp, only to be revived, or to survive, in remnants in Centennial Park. Today, the Lachlan Swamp for Ashton et al. (2013, 69) is "the physical heart and center of Centennial Park" in which water, as the lifeblood of the earth-body, is no longer conveyed to the city.

Centennial Park was the product of the English landscape aesthetic and landscape architecture and gardening imposed on unaesthetic wetlandscapes. In "recounting the wild origins of the vast area that is now verdant parkland," Ashton and his colleagues (2013) also recount the triumphalist history (or sad and sorry story) of Centennial Park from its beginnings in the Lachlan Swamp to its present "undulating lawns" and "ornamental lakes" (Ashton et al. 2013, 23). This transformation took place in a process from swamp, through commons, to park via the imposition of various fads, fashions and styles in park landscape architecture, as well as getting entangled in the politicking of the urban public park movement of the nineteenth century (Schenker 2009; Elborough 2016). The result was "the largest urban park in the southern hemisphere" (Ashton et al. 2013, 11, 27, 52–68).

The "Sydney Common" was set aside in 1811 and:

> shortly after reserved for "the supply of water to the Town of Sydney." Its poor soil was not unlike that of New York's famous Central Park, America's first landscaped urban park [d]esigned by Frederick Law Olmsted' and his partner [. . .] Its site was made up of rocky outcrops, swamps and hillocks [. . .] The Sydney Common also was largely made up of swamp land, sandhills, bluffs and rocky outcrops. (Ashton et al. 2013, 11)

Market gardeners did not find the soil of the Lachlan Swamp poor as they "found marshy portions of the swamps 'uncommonly fertile,'" as they go on to quote somewhat contradictorily (Ashton et al. 2013, 13). Most wetland commons were. And most commons (wetland or dryland) were enclosed into private or public property, such as parks. Ashton et al. (2013, 50 and 52) relate how this occurred in England, and apply it to Sydney Common and Centennial Park. Market gardeners also found the soil of the swamps to the north and south of central Perth to be "uncommonly fertile" (see Ryan and Chen 2020, 15–19).

A similar fate of wetlands ending up being "occupied" by the lawns of public parks and gardens, or by the ornamental lakes, lagoons and ponds in them, befell wetlands in Perth with Third Swamp becoming Hyde Park and in Melbourne with a billabong on the Yarra River becoming the Lagoon in the Melbourne Botanic Gardens and South Melbourne Swamp becoming Albert Park Lake (Giblett 1996, 2013a, 2020c). The draining or filling of wetlands, or stealing water from them, was part and parcel of the project of colonisation. The transformation of wetlands into parks, gardens, lakes, lagoons, and ponds in accordance with the conventions of the European landscape aesthetics and landscape architecture and gardening carried on colonization (for Melbourne's Botanic Gardens, see Giblett 2020c, chapter 5). The British invaders dispossessed the Aboriginal peoples of their lands and stole the water from their streams and wetlands, taking it away to the settlement/city and eventually destroying the swamps, and then transformed what was left into parks and gardens to hide their crime of aquaterracide buried beneath rolling lawns, ornamental lakes, flower beds, and shady trees.

What if Centennial Park had remained a wetland and as a source of water for the city of Sydney? What if it had been wetlandscaped, instead of being parklandscaped into "undulating lawns," "ornamental lakes" and "verdant parkland" in accordance with the conventions of the English landscape aesthetic and landscape architecture and gardening? Wetlandscaping has developed recently as an engineering, aesthetic, and recreational response to water management of relict wetlands in urban areas (as we will see in the following chapter of the present volume in case studies of London and Melbourne). The repressed wetlands also return to haunt the present literally when these area flood and metaphorically in swampy figures for the urban underside (Giblett 2016a). As Marx said, "the tradition of all dead generations weighs like a nightmare on the brains of the living." The tradition of all dead generations that destroyed wetlands by draining or filling them, taking water from them and drying them up, or landscaping them into park, or using them as figures for slums weighs like a nightmare on the brains of the living.

The pioneering, decolonising work of Franz Fanon (1967, 30), in relation to the bush, swamps, and what he calls "indocile" or "hostile" nature demonstrates how colonization was just as much about the draining of swamps as it was about a native population that was economically and politically oppressed by imperial invaders from Europe. In fact, he argues that they were "one and the same thing." What Mbembe (2019) calls "necropolitics" relates to the death of wetlands too, as do genocide to aquaterracide, racism to placism, misogyny to misaquaterrism. Decolonizaton will not be over until the stream, harbor, wetlands, and nature more generally are decolonised. Fanon also critiqued the spatial hierarchy of the archetypal colonial city divided between the "Settler's Town" basking on the hillsides, and the

"Native Town" "wallowing in the mire," as in E. M. Forster's *A Passage to India* (1924), on the mudflats of the Ganges in Chandrapore (Forster [1924] 1936, 9).[1]

Sydney is a case in point with what Turnbull (1999, 219) describes as "rockier land on the western side of the Tank Stream [. . .] allocated to the convicts, sailors and marines, [while] Phillip and his administrative staff occupied the grassier and lusher eastern side of the city beyond the tidal mud flats at the head of the cove." Melbourne is a similar case in point with the first settlement, colonial centre, and now central business district located on the higher and drier northern side overlooked the swampy southern side of lower class suburb of South Melbourne and the initial wetland, later dumping ground, then manufacturing site, and eventual wasteland of Fishermen's Bend (Giblett 2020c, chapters 2 and 3). Further upstream and much later, the social divides on either side of the serpentine river were reversed with Melbourne's working-class suburbs, such as Collingwood, wallowing in the grotesque lower earthly regions of the northern, swampy side of the Yarra River while the upper classes reveled in the salubrious suburbs, such as Toorak, on the southern side (Giblett 2020c, 141–142). Other Australian cities, including Perth (see Giblett 1996, 65 and 74), followed suit.

BACK TO BOTANY BAY

After the Lachlan Swamps dried up, water from the wetlands of Botany Bay "that Phillip had scorned" as unhealthy, was stolen away to Sydney, initially by carters, later by tunnels and pipes (Cathcart 2010, 46). Of Botany Bay Cathcart (2010, 11 and 23) optimistically suggests that:

> today we would celebrate this water-country as wetlands [and indeed we would and do to some extent]. But Phillips and his officers reviled it as marshes, subscribing to an idea as old as Hippocrates, that these were unhealthy places, emanating miasmic gases [. . .] Being an optimist, Phillip foresaw that, in time, man-made channels would drain these malignant marshes.

Phillip was selective in his revulsion as he reviled the malignant and miasmic marshes of Botany Bay enough to avoid them, whereas he revered the benign and salubrious Tank Stream, albeit though it emerged from swampland, enough to try to protect it. Both malignant marshes and benign swampland were wetlands. Both were organs of the living body of the earth, just as malignant cancerous cells and benign noncancerous ones can be part of the same human body.

Draining the "malignant marshes" of Botany Bay by "man-made" channels would make them into benign drylands, and the marshes and their miasma thereby destroyed. The miasmatic theory of disease, especially of malaria (literally "bad air"), caused by marshes and other wetlands prevailed until the discovery in the late nineteenth century that the *Anopheles* mosquito is the vector for malaria before the wide acceptance of "the germ theory of disease" in the late nineteenth century (Cathcart 2010, 35; see Giblett 1996, chapter 5).

The Botany Bay Swamps suffered a similar fate to the Lachlan Swamp with drains eventually being cut through the swamp to increase the yield of water, which only caused the surface to shrink, dry up, and harden (Cathcart 2010, 46). It, too, ceased to be a wetland. Sydney was not born in wetlands, and their watery womb, unlike many other cities, such as Paris, London, Berlin, Hamburg, Venice, St Petersburg, Boston, New York, New Orleans, Washington, Chicago, Toronto, and many others (Giblett 2016a). Yet the infancy of Sydney was sustained by wetlands until they became barren wombs and died in a dry tomb. What if Sydney had used its wetlands sustainably? Would it be facing the water crisis it is facing today? Sydney is a sunny city, but it is also a rainy city receiving about a meter of rain per year, more than twice as much as London and almost double that of Paris (Flannery 1999, 14; Cathcart 2010, 22). It squandered its water resources of nearby wetlands and now water comes from dams much further away, costing a lot more.

Unlike many other cities around the worlds built by, or on, wetlands that drained or filled their wetlands, the settlement of Sydney was built on sandstone, and of sandstone, as Falconer (2010, 21) notes repeatedly, with its "mysteriously porous nature." Porosity might be a metaphor for Sydney's geography, history, and culture, as it was for Walter Benjamin in writing about Naples and its porous stone, spaces, culture, and everyday life (Giblett 2016a, chapter 2).

Tim Flannery (1999, 7–9, 14–15, 35) in the introduction to his anthology of early writings about the birth of Sydney gives a lucid account of the geological history of "The Sandstone City." Yet he does not discuss the nearby Lachlan and Botany Swamps in his introduction (such as the fact that "the base of Botany Bay," "until about 10,000 years ago," was "a swampy sand plain" (Attenbrow 2010, 39), then it became a lagoon and mudflats). Nor does Flannery (ed. 2002) include in his anthology any writers on these swamps, such as those in the archival sources that Cathcart (2010, chapters 1–3) cites and discusses in his three chapters on Sydney. Flannery's anthology includes writings about Sydney from 1770 to the 1890s, thus embracing the period in which these wetlands were described, exploited, and destroyed by settlers. Flannery writes wetlands out of the history of Sydney, just as they were written out of geography by having their waters stolen away and the lifeblood drained from them.

Delia Falconer (2010) in her book on Sydney in the New South Books series on Australian cities also writes wetlands out of the history and geography of Sydney with the exception of the swampland that fed the Tank Stream. By contrast, Sophie Cunningham (2011, 127) in her book on Melbourne in the same series as Falconer's suggests that "to make sense of Melbourne, look to its erratic, brackish wetlands; its muddy, beautiful rivers; its sometimes smelly old lagoons and lakes; and the sudden shock of those moments after heavy rain when the city's cup runneth over" alluding to *Psalm* 23: 5. When there is too much water in the wrong place, the city floods, such as the notorious flood of 1972 when Elizabeth St in central Melbourne flooded and as depicted in a famous photograph (Giblett 2020c, figure 6, 82).

Melbourne's wetlands also fare better than Sydney's do when Flannery (2002, 7–8) in the introduction to his anthology of writings devoted to the birth of Melbourne describes its billabongs and swamps at settlement "as a sort of temperate Kakadu." This description has been taken and up and used by Dean Stewart (2020) in his map of "the ancestral family estates" of the Aboriginal peoples of Melbourne (reproduced inside the front cover of Giblett 2020c). Might not Sydney's wetlands have been a sort of sub-tropical Kakadu at settlement? Might not Perth's wetlands on the Swan Coastal Plain in Western Australia at settlement have been a sort of Mediterranean climatic Kakadu of the south-west? In the introduction to his anthology of early writings about Melbourne Flannery goes on to quote (albeit selectively) George Gordon McCrae's lyrical and elegiac evocation of the Blue Lake (Batman's or West Melbourne Swamp; see the following chapter of the present volume). Flannery (ed. 2002) also includes in his anthology devoted to Melbourne many writers on these and other wetlands, including those in the archival sources that Cathcart (2010, chapter 5) cites in his chapter on Melbourne.

Presumably Sydney's wetlands were formed when, millions of years ago, one river's "vast fossilized floodplain" (Flannery 1999, 9–10), together with another river, millions of years later, created the site of Sydney. As Falconer (2010, 35) puts it succinctly, "the whole of metropolitan Sydney is built on the great bed of a prehistoric floodplain." The Lachlan and Botany Swamps are relic remnants of the prehistoric floodplain and the rivers. They were present at the geological and colonial birth and infancy of Sydney. They suffered from the sudden infant death syndrome borne out of the illicit intercourse between colonial settlements/cities and native swamps. The settlement destroyed the Lachlan and Botany Bay Swamps by taking their water away from them in a visible process above ground and then in an invisible process underground, until the impacts on the surface, in the wetlands, became visible and they ceased to be wetlands, and were largely forgotten. The nearby wetlands of Sydney were well and truly oppressed and repressed. The birth of cities by or near wetlands and their infancy spent exploiting them meant the death of those wetlands. Sydney was an exception to many cities in being built on sandstone, but it was not an exception to such cities as Paris, London, Venice, Berlin, Hamburg, St. Petersburg, New York, Boston, New Orleans, Toronto, Washington, Chicago, and Melbourne that destroyed their wetlands, often with devastating consequences when cities flood.

By another irony of history, the name of Lachlan Swamp that had been applied at the time of the first settlement at Sydney Cove to a nearby wetland, and now to the dispossessed wetlands of Centennial Park in Sydney, was taken and applied shortly thereafter to the swamps on the Lachlan River in inland New South Wales. The death knell of one Lachlan Swamp was sounded in the naming of another Lachlan Swamp, whose death knell was sounded too, as much of it has now been drained (Cathcart, 2010, 91). John Oxley led an exploration expedition in 1817 to discover where the Lachlan River went, whether it was a river that flowed into another river (it does, into the Murrumbidgee) or into the fabled inland sea (it doesn't—that body of water ceased to exist millions of years ago; see Flannery 1999, 10).

WEIRD AUSTRALIAN MELANCHOLY

Oxley (cited by Cathcart 2010, 91; see also 94) described how he and his party found that the Lachlan River went into "sickly marshes and unhealthy places" that were associated not only with "death" in conformity with the miasmatic theory of physical illness, but also with "melancholy" in accordance with the miasmatic theory of mental illness. Oxley, for Cathcart (2010, 91), "had gone in search of the river of life, only to have his heart broken by the arid plains of inland New South Wales." These plains can be dry in drought, but the marshes can be, or were wetlands, and the soil fertile—water-bodies of life. As Cathcart says earlier on the same page, "today [. . .] much of the wetlands have been drained. The surrounding country is occupied [that word again] by the prime sheep-and-wheat country around Condolbin and Forbes." Prime pastoral and agricultural country is the product of fertile wetlands. A similar story can be told throughout many parts of Australia. Might not the swamps and marshes on the Lachlan and Macquarie Rivers have been the Kakadu of inland New South Wales at settlement? Recently 80,000 hectares of the Lachlan Murrumbidgee wetlands have been dubbed the "Kakadu of the south" and "handed back" to Aboriginal owners (Ruzicka 2020). Nature Conservancy Australia (2020) noted that "for thousands of years the First Australians in this area made interventions to boost the productivity of their Country—enhancing fish and bird stocks, and vegetation growth." Ten percent of the Macquarie Marshes have been a Nature Reserve since 1971 and a

Ramsar Convention Wetland of International Importance since 1986 (NSW National Parks and Wildlife Service, n.d.).

Oxley, and later Charles Sturt (cited by Cathcart 2010, 110), reviled these marshes as melancholic, subscribing to an idea as old as Robert Burton and John Bunyan that miasmatic effluvia emanating from marshes caused melancholy (as we saw in chapter 2). John Bunyan's *Pilgrim's Progress* was first published in 1678 (Bunyan 2008). It has been described by Louis Marin (cited by Giblett 1996, 166–167) as "the most influential religious book ever composed in the English language." It is also the second most published book in the English language after the Bible, and so it is ahead of other classics in English, such as the complete works of Shakespeare. Part of its influence has been to convince readers that a perfectly good wetland is not what it seems and is, but is, in fact, a slough of despond that would drag the unwitting pilgrim down into the depths of despondency and melancholy, and even into despair and suicide. Bunyan's "Slough of Despond" was following in the despondent muddy footsteps of Robert Burton's *Anatomy of Melancholy* published in 1621 and republished four times within the succeeding seventeen years (Burton 1932, 224–225, 239). In 1789 Watkin Tench was an early uptaker and perpetrator of melancholy in relation to the Australian landscape (cited by Cathcart 2010, 27). Instances of the miasmatic theory of mental and physical illnesses of melancholia and malaria could be enumerated almost ad infinitum, certainly ad nauseam (see Giblett 1996, chapters 5 and 7).

Oxley for Cathcart (2010, 89–98) was a key figure of what he calls "the Great Australian Melancholy" associated with "the Great Swamp" of the Lachlan Swamp and Macquarie Marshes. Associating melancholy with muddy marshes, despondent sloughs, and slimy swamps (great or small) is an idea as old as John Bunyan and Robert Burton in the seventeenth century, William Byrd in the eighteenth century who named the "Great Dismal Swamp" in the United States and Theodore Roosevelt in the twentieth century who saw all swamps as dismal (Giblett 1996).

Melancholy becomes the key-tone for Cathcart with which Australian explorers, such as Oxley and Sturt, and writers, such as Marcus Clarke and his "weird melancholy," regarded the Australian bush and the outback. Melancholy for Cathcart (2010, 151; see also 175) is associated with "the mythology of the great Australian bush" and he cites some instances in early twentieth-century Australian poetry on the following page. It was the symptom of what Cathcart (2010, chapter 11) in the following chapter diagnoses, and calls, "necronationalism," a national, and nationalist, obsession with death that was manifested ultimately in the ANZAC myth of Gallipoli (Cathcart 2010, 163). "The melancholy of necronationalism" for Cathcart (2010, 213) is a persistent tone in Australian culture. As with Mbembe's (2019) necropolitics, Cathcart's necronationalism relates to the death of wetlands too, as do genocide to aquaterracide, racism to placism, misogyny to misaquaterrism, but neither author considers this connection.

In the same year as Cathcart published *Water Dreamers*, Delia Falconer (2010, 2) published her book about the city of Sydney in which she suggests that "its fundamental temperament is melancholy." This is hardly surprising given the number of wetlands it has destroyed and lost. It has also had what she calls an "interest in the bodily humors like bile and phlegm" (Falconer 2010, 5). Bile is melancholy in the mainstream western theory of the elements and humors caused or created by the black waters of wetlands with their rising miasma blocking out the sun and turning it into the black sun of melancholy (see figure 3.1, this book). Melancholy is phlegmy and slimy in the mainstream western theory of the elements and humors devised by ancient Greek philosophers and later formulated in the Elizabethan "world picture" (see Giblett 1996, 160–161). It is a fitting humor and temperament for a city such as Sydney with so much water in its harbor and off its coast, and a city founded near wetlands, now lost. It is also a fitting humor and temperament for a capitalist city as Eagleton (1986, 41) argues that "melancholia is an appropriate neurosis for a profit-based society."

For Falconer (2010, 36) "it would be difficult to find a city more permeated by water." This is the case in the present and even more so in the past. Sydney for Falconer (2010, 21) is a city haunted by its past. She goes on to suggest that "Sydney is not so much full of ghosts, as absences" (22), though she acknowledges later "the ghost creeks" (p. 35) of the city. Melbourne could be called a "city of ghost swamps" whose spectral remnants of swamps haunt the city today (Giblett 2020c, part I). Sydney is a city today haunted by its absent swamps and ghost creeks. Sydney is like Venice, as Kenneth Slessor noted (cited by Falconer 2010, 36, 54). Sydney is the Venice of the south, whereas Melbourne is the Paris of the south (Giblett 2020c, chapter 6). Both Paris and Venice are cities built on or beside wetlands, like Sydney and Melbourne (for Paris and Venice, see Giblett 2016a, chapters 3 and 5).

Associating melancholy with mourning is an idea that goes back at least to the German mourning plays of the sixteenth and seventeenth centuries made famous by Walter Benjamin in his 1925 study of them (as we saw in chapter 2; see also Benjamin 1977, 2019). It is also an idea present in Sigmund Freud's famous essay on mourning and melancholia of 1917 (as we also saw in chapter 2; see also Freud 1984, 2005). In mourning, the world is experienced as loss, whereas in melancholia the self is experienced as lost. The lost wetlands destroyed by cities continue to exist, now as part of the dejected subject of the citizen who maintains "a relationship with an absent lost object" (as Ferber puts it and as cited in chapter 2), just as the subject maintains a relationship with absent lost wetlands by being oblivious to them or by retelling their stories, by indicating their presence in the past on maps and their absence in the present on maps too (as Cathcart 2010, 95 does in this map that shows the Lachlan Swamp and Macquarie Marshes in Oxley's time with a standard cartographic symbol for marsh and swamp).

As wetlands are increasingly lost from the world and are lost as an object of love which nourishes life, both mourning and melancholia are experienced and exercised in relation to them. The world should be in mourning for the loss of its wetlands which gave it life and nourished it, but instead of being in mourning and regarding the world as losing its wetlands becoming empty or "bereft" of them, in Gerard Manley Hopkins (cited by Giblett 2019d, 24) apt word, it experiences this loss as a melancholic loss of its own ego, its own selfhood and sense of identity (as we saw in chapter 2).

MASCULINIZED MATESHIP AGAINST FEMINIZED BUSH

Melancholia is a specific masculine affliction and view of the self in relation to the Australian bush and the outback gendered as feminine. It sets up and maintains the solidarity of masculinized mateship against feminized bush. Mateship was born of masculine solidarity against the bush. Flora Eldershaw and Marjorie Barnard reflected singly, jointly (as M. Barnard Eldershaw) and somewhat critically on the phenomenon of Australian mateship in the 1930s, 1940s and 1950s. Mateship for them is "the solidarity of men against the bush" (Eldershaw 1939, 52), or elsewhere for Eldershaw (1952, 217), against nature. Mateship is the formation of patriarchal, or perhaps more precisely filiarchal, solidarity, the brotherhood of white men, of southern sons, of the "coming man" of colonial modernity, against the threats posed by the feminized black bush and its (or "her") Black people.

Flora Eldershaw and Marjorie Barnard (Eldershaw 1939, 223) advise that "the pioneer must take the offensive against the land [. . .] Defeat may be final but victory never is." The pioneer must take arms against the sea of troubles posed by the bush for by not opposing them they will end him. For them, "the bush is like the sea, it has no paths, always the same, unchanging, the silence made visible" (Eldershaw 1939, 13; see also 53 and 61). Unlike wilderness that made the mute land speak, the bush made silence visible. The bush could be represented in painting, but wilderness could never be—it could only be gestured toward. The wilderness could be represented in writing but the bush could never be—it could only be defined. Yet mateship is not just passive male bonding among men to produce the ties of hom(m)oerotic friendship but also active masculine colonisation of the feminized bush. Eldershaw (1952, 217) argues that "men must assert themselves against the lonely immensities and scrawl something in the monotone."

Cathcart (2010, 139–140) bypasses Barnard and Eldershaw and dismisses the commonplace in some feminist academic writing about Australia over the past thirty years that critiques the mainstream masculinist gendering of "the bush" as feminine. His specific target is Kay Schaffer's (1988) *Women and the Bush*, clearly a groundbreaking study when it was first published and a broadly influential study ever since, including very recently.[2] Cathcart does not address Schaffer's central argument that metaphor is women's burden (Schaffer 1988, xii; see also Giblett 1996, 47 and 51, n.16). Instead, in a sweeping gesture and gross generalization, Cathcart (2010, 139–140) asserts that gendering "the bush" as feminine has no basis in nineteenth-century explorers' journals.

Yet in 1836 Major Mitchell unmistakably gendered and sexualized "Australia Felix" of Western Victoria is his exploration journal (as we saw in chapter 4 of the present volume). Perhaps needless to say Cathcart does not cite Mitchell's description of "Australia Felix" in his extensive use of Mitchell's work throughout *The Water Dreamers* as it flies in the face of his assertion (nor does Schaffer for that matter, though it supports her argument). Cathcart may contend in his own defense that he is not writing about water in Western Victoria so he could ignore Mitchell's description of "Australia Felix."

Some twentieth-century writing about the Australian bush is no different to the nineteenth century explorers when it comes to gendering "the bush" (as Cathcart does not show). Cathcart cites D. H. Lawrence's work in passing, including his 1923 novel *Kangaroo* on one occasion, but does not note that in it the Australian bush is split between a fear of, and fascination with, the Paleolithic Great Goddess/Mother of the swamp world, and a desire for and a fantasy of a young, virginal nature:

> And they ran on bridges over two arms of water from the sea, and they saw what looked like a long lake with wooded shores and bungalows: a bit like Lake Como, but oh, so unlike. That curious sombreness of Australia, the sense of oldness, with the forms all worn down low and blunt, squat. The squat–seeming earth. (Lawrence 1950, 86)

The landscape is at first measured against the European norm and found to be different. It is then described in terms of the intractable old Great Goddess/Mother (she is even figured in terms of the squat figurines depicting her).

The Great Goddess/Mother earth is contrasted with the lithe, virginal bush (she is even figured as a young Aboriginal woman in statuesque pose in a static tableau which may say more about the fact that the scene is being viewed from a train than anything profound about the landscape itself):

> And then they ran at last into real country, rather rocky, dark old rocks, and sombre bush with its different pale–stemmed dull–leaved gum–trees standing graceful, and various healthy–looking undergrowth, and great spiky things like yuccas. As they turned south they saw tree–ferns standing on one knobbly leg among the gums, and among the rocks ordinary ferns and small bushes spreading in glades and up sharp hill–slopes. It was virgin bush, and as if unvisited, lost, sombre, with plenty of space, yet spreading grey for miles and miles, in a hollow towards the west. Far in the west, the sky having suddenly cleared, they saw the magical range of the Blue Mountains. And all this hoary space of bush between. (Lawrence 1950, 86–87)

The Australian landscape for Lawrence is split between the hoary old Great Goddess/Mother earth and the young virgin bush.

The Australian landscape could be beautiful for Lawrence, and even more:

> The strange, as it were, *invisible* beauty of Australia, which is undeniably there, but which seems to lurk just beyond the range of our white vision. You feel you can't *see*—as if your eyes hadn't the vision in them to correspond with the outside landscape. For the landscape is so unimpressive, like a face with little or no features, a dark face. It is so aboriginal, out of our ken, and it hangs back so aloof . . . And yet when you don't have the feeling of ugliness or monotony, in landscape or in nigger you get a sense of subtle, remote, *formless* beauty more poignant than anything ever experienced before. (Lawrence 1950, 87)

In other words, or in one word, you can experience the sublime. Yet perhaps this was the minority view of a sophisticated European (albeit of working class origins) of the 1930s not shared by the majority of Australians who saw the landscape in more prosaic terms.

Lawrence is following in Major Mitchell's footsteps with his split between the soft, flowery and virgin land and the dry and barren desert (as we saw in chapter 4). Similarly (as we also saw in chapter 5), Woolnough masculinises mountains and feminises and splits salt lakes between the good, young, beautiful white virgin on the surface and the bad, dried up, uncanny old dame in the depths beneath the surface as a defense against the horrors of what he sees, but does not name, as "Western Australia Execratus." Cathcart also overlooks or ignores the affirmative gendering of wetlands, specifically the association between the Great Mother or Great Goddess and the swamp, an idea as old as J. J. Bachofen in the nineteenth century (Giblett 1996, 34–36), and taken up by Walter Benjamin (Giblett 2020c, chapter 6) and Erich Neumann in the twentieth century (albeit inflected with misogyny in the latter case; see Giblett 1996, 86–90, 145–147, 175–176).

The gendering and sexualizing of Australian drylands, wetlands and water is prominent in Mary Durack's *Kings in Grass Castles* published in 1959 and mentioned briefly by Cathcart (2010, 279, n.38). It is a settler family saga in which her squatter forebears drove their stock across the country, commandeering Aboriginal waterholes, "devouring the[ir] grasslands," as Cathcart (2010, 121) puts it of squatters in general, and trampling their billabongs and other wetlands into useless mud. Devouring is an act of oral sadism as I have argued elsewhere and a psychogeopathology as I have diagnosed in relation to *Kings in Grass Castles* (see Giblett 2019d, chapter 6). *Kings in Grass Castles* is also a boy's own masculinist adventure story and wet day dream in fine imperial fashion in which the good, soft land is figured, and fantasized, as a passively supine female body laid open to and decorated for the penetrating gaze of the epic hero along similar lines to Major Mitchell. The bad, dry land is figured, and fantasized as the old, dry dug of the bad breast that contrasts with the deep and wide expanse of the good breast, the Ord River, and with the lush pasture of the wet season figured as a blowsy, slightly tipsy, middle-aged outback mother who is both sexually fascinating and physically repugnant to the randy drovers (see Giblett 2019d, chapter 6).

From the beginnings of the first settlement in Australia in Sydney in 1788, through subsequent settlements of Perth founded in 1829 and Melbourne founded in 1835, to the beginnings of the twenty-first century, Australian wetlands have had a rough deal in fact, fiction, and nonfiction. Today we should mourn the loss and absence of most of the Australian wetlands that were present at the time of the British invasion, commemorate their lives in story, history, memorials, interpretation, and other arts, and celebrate and conserve our remaining wetlands still present today. They are water-bodies of life. As are we human beings. How can one water-body live without the other? Water bodies matter!

NOTES

1. All these environmental aspects of Fanon's work are unfortunately neglected or overlooked by Mbembe in his otherwise exemplary and powerful study entitled *Necropolitics* whose avowed basis is Fanon's work (Mbembe 2019, 2) and recently in his equally exemplary and powerful collection of essays on decolonization, *Out of the Dark Night* (Mbembe 2021). For a recent discussion of the decolonization of nature (with a brief nod to Fanon) see Obeng-Odoom (2021).

2. See, for instance, Potter (2019, vii) who dedicates her book to Schaffer; see also Giblett (2011, chapter 6).

Chapter 7

Water in Urban Waterscapes and Wetlands in London and Melbourne

Water is vital for life on earth. Drinking water sustains life. Plants and animals (including humans) depend on it and can't survive for long without it. Water is precious and dangerous. Wastewater disposes of dead, or deadly, matter. Water is both life-giving and death-dealing. Too much of it in floods and too little of it in droughts are both deadly. Climate catastrophes, such as storm surges and rising sea levels, in the age of global heating are a present and imminent danger of too much water in the wrong place, at the wrong time, to people, places, plants, animals, and livelihoods in low-lying coastal and riverine areas. Still water contained in rivers, streams, ponds, and lakes is beautiful or picturesque to look at depending on the scale or point of view.

Clean water in which to bathe and swim is cleansing and refreshing for the human body. Swimming water with surf, rips, and snags are hidden deadly dangers in oceans and rivers. Salt water in mountainous seas and vast oceans is sublime and terrifying in immensity and power. Saline water on the farm is deadly, killing fertile land. Stagnant water in slimy swamps, muddy marshes and other wetlands is uncanny, but also fertile and fecund. Flowing water in raging torrents, then spreading out over floodplains, inundating houses, shops, towns, is destructive and soul-destroying. Drying, drought-ravaged paddocks and rangelands equally so. Fresh water is beautiful and soothing to drink; wastewater can be a carrier of disease. City and country are linked through water and in catchments dams and pipes.

Water is seen, carried, transported, disposed of, designed for, and so on, in and for two cities in four different paradigms in recent work on Melbourne and London in the disciplines, discourses and practices of: urban planning (Dovey and Jones 2018; Jones 2018); landscape architecture (Bertram and Murphy 2019); urban geography (Gandy 2014); and wetland cultural studies in the transdisciplinary environmental humanities (Giblett 1996, 2009, 2016a, 2020c).

Water in urban planning, especially when it is embanked in waterfronts with hard edges and a strict demarcation between land and water, is largely an object of visual appreciation and recreational consumption. Water in landscape architecture, especially when it comes to former wetlands, now wastelands of former industrial sites, is largely an object of "water sensitive urban design" in the rehabilitation of lost wetlands and the creation of artificial wetlands that not only provide ecological services (such as flood mitigation) and recreational amenities in an aesthetically appealing urban environment, but also echo the historic swamps and reinstate a remnant of them.

Water in urban geography, especially when it comes to cities and the delivery of potable water in pipes and the disposal of wastewater in sewers, is largely an object of engineering and neo-colonial "management." When it comes to water in urban landscapes, especially in reclaimed wetlands, urban geography takes a similar approach to landscape architecture in re-reclaiming wetlands.

Water in wetland cultural studies in the transdisciplinary environmental humanities, especially when it comes to wetlands, is largely an abject of anti-aesthetic appreciation in the uncanny. When it comes to cities and the country, water in the environmental humanities is a matter of conservation in bioregional catchments manifesting in the bodies and flows of water as the lifeblood of the earth body in rivers and wetlands of home habitats of the living earth. The environmental humanities take a transdisciplinary historical, geographical, philosophical, and cultural approach to water and wetlands.

Wetlands—swamps, marshes, mires, morasses, bogs, lagoons, sloughs, shallow lakes and estuaries, and so on—have been seen in patriarchal western culture as places of darkness, disease and death, horror and the uncanny, melancholy and the monstrous—in short, as dead black waters (see Giblett 1996). Elsewhere I have proposed seeing them in indigenous cultural terms as living black waters (Giblett 2007; 2009, chapter 11). They are not necessarily the most pleasant of places, but they are places of new life being born out of the old and from death. Instead of seeing wetlands and other such water bodies as black waters linked to death, let's see them as blood-red concurrent with both life and death. The value-laden, color-coding of water—white is good, black is bad—flows through into other accounts of water—light blue is good, dark blue is dangerous, brown is bad (as we will see below in relation to Melbourne). Red is an anomaly as it, like blood and water, is good and bad, or bad like rust (as we will also see below in relation to Melbourne).

Rather than seeing water in white terms as black or white, in black and white terms, I propose seeing it as blood and blood-red pointing to both life and death. Instead of seeing the waters of the earth as a system of canals for irrigation and transportation, let's see them as the lifeblood of the body of the earth. Let's see rivers as a digestive tract and wetlands as organs, such as the kidneys, liver, and placenta with rivers and wetlands living together as part of one whole body. Let's deconstruct and decolonise the white cultural coding of water as black or white and take on board an indigenous color-coding of water as red, as water, like blood, that is both life-giving and death-dealing, the lifeblood, of the great, wide brown body of land of Australia. The human body and the earth would then be intertwined in one living being—the body of the earth (Giblett 1996, 2007, 2008a, 2009, 2011, 2014, 2016a, b).

Wetlands and cities have had a fraught, if not inimical, relationship: where the city is now, there the wetland was once; where the wetland is now on the outskirts of the city, the city, or its suburbs, soon will be; where the restored, rehabilitated, or artificial wetland is in the city, a wetland or wasteland once was (Giblett 2016b). Many major cities were built on or beside wetlands, including all the major, iconic cities of modernity in Europe and North America such as Paris, London, Venice, Berlin, Hamburg, St Petersburg, New York, Boston, New Orleans, Toronto, Washington, and Chicago. These cities drained or filled wetlands (Giblett 2016b). Many major cities in Australia were also built on or beside wetlands and drained or filled them too. These cities include Sydney (as we saw in the previous chapter) Melbourne (Bertram and Murphy 2019; Giblett 2016, 2020c) and Perth (Giblett 1996, chapter 3; 2013a, chapter 15; Martin and London 2019; Brady and Murray 2020; Ryan, Brady and Kueh 2020).

Wetlands are vital for life on earth, including human and nonhuman life. The leading intergovernmental agency on wetlands states that:

> they are among the world's most productive environments; cradles of biological diversity that provide the water and productivity upon which countless species of plants and animals depend for survival. Wetlands are indispensable for the countless benefits or "ecosystem services" that they provide humanity, ranging from freshwater supply, food and building materials, and biodiversity, to flood control, groundwater recharge, and climate change mitigation. Yet study after study demonstrates that wetland area[s] and [their] quality continue to decline in most regions of the world. As a result, the ecosystem services that wetlands provide to people are compromised. (Ramsar Convention Bureau, n.d.)

Yet more than the mere providers of "ecosystem services," wetlands are habitats for plants and animals, and homes for people. They are also principally under threat on the outskirts of cities where they are drained and filled to create sites for more homes for more people.

A wetland cultural studies' view of water in cities and the country and of wetlands everywhere is leading to the possibility and prospect of wetlandscape architecture following in, and deviating from, the footsteps of Frederick Law Olmsted, the founder of American landscape architecture, beginning with his pioneering work in his design for the Back Bay Fens in Boston, arguably the world's first artificial, or constructed, wetland. Olmsted's pioneering wetlandscape in Boston, however, was overshadowed by his famous work on the wet and dry park landscape of Central Park in New York in which wetlands were filled or drained or aestheticized into lakes (and as we have seen with Sydney and Centennial Park in the previous chapter and as occurred with Melbourne and the Botanic Gardens; see Giblett 2020c, chapter 7 for Melbourne). Olmsted did not regard his design for Back Bay Fens as a park, and indeed it isn't a park at all as parks are gardens, lawns, and lakes predicated on a hard and fast divide between land and water with water in lakes or ponds, grass in lawns, trees in avenues, and bushes in garden beds all combined to produce pleasing pastoral prospects. Wetlands, by contrast, are in transition between land and water. Wetlandscape architecture constructs artificial wetlands where land and water are in transition from one to the other. Wetlandscapes should be regarded as legitimate places of landscape design and should not be seen as second-class citizens to first-class park landscapes (Giblett 2016, chapters 8 and 9 for Olmsted, New York and Boston).

URBAN CHOREOGRAPHY

Melbourne is a case in point as a couple of recent books show. *Urban Choreography* is the title of a recent account of the development of Melbourne's Central Business District (CBD) from the mid-1980s. In their introduction to this collection of essays, Ken Dovey and Ron Jones (2018, 9) announce that "having once turned its back on the water, Melbourne has now embraced the Yarra River and Port Phillip Bay, becoming a waterfront city" after having embanked the Yarra. From embanking to embracing indicates Melbourne's changing attitude to the Yarra River (but not to its wetlands as we will see shortly). Melbourne is like Toronto in this regard, and other cities no doubt (for Toronto, see Giblett 2014, chapter 6; 2016a, chapter 11). Before it was "a waterfront city" with hard, embanked edges, it was a wetland city with soft, muddy liminal zones, again like Toronto and many other cities (Giblett 2016a). The presence of wetlands on the "banks" of the Yarra River in the past is well-documented (and their absence in the present mourned; see Giblett 2016, 2020c).

Wetlands do not rate a mention as part of "the water," or as part of the past of the present "waterfront" of Melbourne in Dovey's and Jones's introduction, or anywhere else in *Urban Choreography* for that matter, except for the much later brief mention of "the marshy Flinders Streets lagoons"

(Jones 2018, 124). Jones's chapter is entitled, "Melbourne, Sung as it were a New Song," but he is singing Melbourne into being from the old colonial song sheet of cities built on drained and forgotten wetlands, and not from the new conservation song sheets of wetlandscape design in landscape architecture and wetland cultural studies in the transdisciplinary environmental humanities.

In their defense, Dovey and Jones might claim that they are only concerned with the development of Melbourne from 1985 onwards (as proclaimed in the subtitle of *Urban Choreography*), but the city cannot be considered in isolation from its previous history of settlement in the swamps and marshy delta of the Yarra and Maribyrnong Rivers, beginning with the European discovery of its site of "swampy country" in 1802–1803 and settlement in 1835, its pre-settlement history in Aboriginal society and culture over tens of thousands of years (Massola 1968), its pre-history in geomorphology and hydrogeology over millions of years and the engineering of its site by filling wetlands and rerouting rivers since settlement (Hills 1975; Bertram and Murphy 2019, esp. 102–105; Giblett 2016, 2020c).

This Aboriginal country, including the Yarra and Maribyrnong Rivers and their swampy deltas, as well as the Moonee Ponds Creek, affected the location, layout, and development of the city. The city of Melbourne is where it is because the Yarra provided potable water for settlers and their stock, and because the wetlands provided feed for stock. The wetlands were an obstacle to development that was overcome by being drained and/or drained after being polluted as waste dumps. Melbourne was founded among wetlands that were drained and filled. The lost wetlands still haunt the city as a virtual presence, as "a city of ghost swamps" (Giblett 2020c, part I, chapters 2–5).

Just as Paul Carter in his account of the history of Melbourne as "a grid-plan town" in *The Road to Botany Bay* ignores or overlooks the wetlands that were on the edges of the "Hoddle Grid" of Melbourne's CBD (Carter 1987, 202–209; see Giblett 2020c, chapter 5), so *Urban Choreography* largely ignores or overlooks the wetlands that were on the "waterfront." Similarly, just as Carter abstracts place into space in the making of *Nearamnew,* his inscription in stone of the history of the colony of Victoria at Federation Square located above "the [former] marshy Flinders Street lagoons" and in his account of its design and construction in *Mythform* (Carter 2005; see Giblett 2016, 2020c, chapter 5), so *Urban Choreography* (2018) abstracts wetlands and river into water, a resource to be exploited, including for its views and for recreation on or beside it. Just as *The Road to Botany Bay* and *Mythform* ignore or overlook the Aboriginal story of Melbourne and its wetlands as places of sustenance for animals, plants, and people (Giblett 2016, 2020c, chapter 5), so does *Urban Choreography*. Melbourne has not embraced its wetlands past and present (as the writers of *Urban Choreography* demonstrate), and as it continues to turn its backside on the wetlands by ignoring, destroying, or polluting them. Just as *The Road to Botany Bay* and *Mythform* are neo-colonial, so *Urban Choreography* is the triumphal neo-colonial dance of death on the grave of Melbourne's lost wetlands.

The map of "Central Melbourne" on the facing page to the comments of Dovey and Jones (2018, 8) quoted above in their introduction to *Urban Choreography* shows an unnamed Yarra River as a mythologized solid blue line of clean water wending its way smoothly through the city (when the Yarra is in fact a sluggish river with brown-colored, muddy water that famously flows upside down; see Giblett 2020c, chapter 8). This moralized and value-laden color-coding of water—blue or white is good, black or brown is bad—flows through other accounts of Melbourne waters (as we will see below). The lost wetlands on the periphery of central Melbourne are invisible. Recent plans for the redevelopment of Federation Square will result in turning the face of the city toward the Yarra River and giving greater access to the Yarra River. Whether this constitutes embracing and celebrating the lost and found wetlands of Melbourne is another question. The answer is "no" if *Urban Choreography* (Dovey and Jones 2018; Jones 2018) is anything to go by.

The Victorian Government has been in public consultation about the future of Federation Square. I have made a submission suggesting the construction of a diorama of maps and photographs of the wetlands that were on the site and elsewhere in Melbourne. Unlike other cities, such as Perth (Western Australia) which has memorials to its lost wetlands, and even an artificial perched wetland outside the Art Gallery of Western Australia in which kids can play, Melbourne has none.

Also unlike other cities, such as Perth, which held an exhibition in its Town Hall in 2014 with maps and photographs of its lost wetlands, Melbourne had had no similar exhibitions until the Royal Historical Society of Victoria's (2020) recent exhibition devoted to maps and photos about West Melbourne (or Batman's) Swamp. The materials for the Perth exhibition, "Reimagining Perth's Lost Wetlands," are now available on the website of the Western Australian Museum (2014). The Victoria Museum does not have anything similar on its website about Melbourne's lost wetlands.

I can't help wondering facetiously about the title for the Melbourne exhibition, "The Swamp Vanishes," in relation to the film, *The Lady Vanishes*, the classic 1938 British mystery thriller film directed by Alfred Hitchcock. Swamps are often feminised in either/or terms, either as the Great Goddess/Mother who is both life-giving and death-dealing, or in misogynist terms as an initially seductive, but ultimately monstrous and orally sadistic (man-eater) femme fatale. Perhaps it is no surprise that West Melbourne (or Batman's) Swamp has been feminised as the "Blue Lake," that it has vanished and that its disappearance is now commemorated in an exhibition with chilling overtones of a mystery thriller. This is also especially the case in the case of the swamp lady who vanished. In 1952 "the swamp lady" vanished (Shears 2008). This became a notorious cold case. The question of who murdered the swamp lady is still an open question, as is the question of who murdered the swamp.

THE BLUENESS OF THE BLUE LAKE

The early settler George Gordon McCrae (1912, 117) described West Melbourne (or Batman's) Swamp as a blue lake when he looked back nostalgically and dreamed poetically how, in the 1840s, it was "a beautiful blue lake [. . .] a real lake, intensely blue, nearly oval, and full of the clearest salt water; but this by no means deep." The name "Blue Lake" has kind of stuck and become immortalised, thanks largely to a recent book about it (Sornig 2018). It sounds much more attractive and romantic than the humdrum and prosaic "West Melbourne (or Batman's) Swamp" based on a compass point in relation to a city centre and a syphilitic settler and co-founder of the settlement that grew into a city (Giblett 2020c, 32).

The Blue Lake resonates with the early nineteenth-century German writer "Novalis" in a posthumously published romance in which the blue flower was "the symbol of romantic longing, the realization of the poet's dream, the union of the dream world and the real world" (Thayer 1954; see also Pastoreau 2001, 138). McCrae's "Blue Lake" was such a union of the two worlds for him. The blue flower is in short what Warwick Mules (email to Giblett April 2, 2020) calls "the impossible real."

The Blue Lake also resonates with the 1930 classic German film *The Blue Angel* starring Marlene Dietrich as a femme fatale cabaret singer. The femme fatale embodies in her seductive, but dangerous, watery body the impossible real. Water bodies, such as the Blue Lake, embody the impossible real for McCrae. The West Melbourne (or Batman's) Swamp was a beautiful body of blue water and a dangerous femme fatale singing a siren song that seduced the romantic poet in McCrae into a wet daydream that ends in a nightmare (or bad daydream) for him seventy years later (as we will see shortly).

The Blue Lake resonates also with another classic German film, *The Blue Light*, produced in 1932, directed by, and starring Leni Riefenstahl in the lead role in her first film who, due to her "feral strangeness," is regarded as a "witch" by the local villagers. The story is a modern, adult fairy tale in the tradition of E. T. A. Hoffmann's uncanny stories, such as "The Mines of Falun" (see Giblett 2019b, 7, 96–99). The blue light in the mountains emitted by cool crystals lures young men to their deaths, but "the wild child" is made to blame for these deaths. The location of the crystals is betrayed by a painter from the city. The greedy villagers steal the crystals in an allegory of city versus country, voracious mining versus magic mountain, not unlike the stealing of the Blue Lake by greedy settlers.

The Blue Lake's fulfilment of the poet's romantic wet day dream had become a modern bad daydream or nightmare seventy years later when McCrae (1912, 117) advises his contemporary readers that "you may search for it in vain to-day among the mud, scrap-iron, broken bottles, and all sorts of red-rusty railway *débris*—the evidences of an exigent and remorseless modern civilization." This value-laden color-coding prefers good blue lake to bad red rust (and wouldn't we all? But by then it is too late and McCrae mourns the swamp he has lost). Blue and melancholy were associated in medieval poetry with "the play on 'ancolie' (a blue flower) and 'mélancolie,'" as Pastoreau (2001, 138–140) points out. Indeed, Rebecca Solnit (2005, 122) traces how "the term *blue* comes from an old English word for melancholy or for sadness, blue moods, blue devils, the blues, first tracked to 1555 in my etymological dictionary."

Blue and mourning became associated in the twelfth century CE with the Virgin Mary and her mourning the death of her son Jesus Christ, as Pastoreau (2001, 50–53) relates and illustrates. Mourning and melancholy were closely associated by Sigmund Freud (2005) in his essay of 1917. Moreover, mourning, melancholy, marshes, and other wetlands, such as the Bunyanesque "slough of despond," have also been associated with each other (as we saw in chapter 2). The Blue Lake was ruined in muddy, red rust and McCrae was left bewildered and bereft. Industrial capitalism transformed the romantic blue lake into modern mud and red rust, a typical fate for wetlands on the urban fringe to become rubbish dumps with the detritus of industrial and consumer capitalism, the endpoint of capitalist production and consumption in waste lands.

The Blue Lake also has dark overtones of *The Blue Dahlia,* the 1946 American film noir (literally "black film") based on Raymond Chandler's original screenplay, and *The Blue Gardenia*, the iconic 1953 film noir directed by Fritz Lang of *Metropolis* fame and starring Anne Baxter as the femme fatale who was billed sensationally in one poster for the film as "there was nothing lily-white about her." The femme fatale is a "deflowered" virgin (or a de-blue-flowered "virgin," another instance of "the impossible real"). What strange blue flowers grew in the black films in the garden of the land of the free, the home of the brave!

Unfortunately, Pastoreau (2001) does not consider in his cultural history of the color blue the proliferation of the use of blue in the titles of these films from the 1930s to the 1950s, nor the association between blue and the femme fatale. In his defense, he announces upfront in the introduction that *Blue: The*

History of a Color is "not meant as a complete history of [the] color [blue] in Western culture" (Pastoreau 2001, 11). Indeed, it is not. Whether the voluminous feminist literature on femme fatales and films noir considers these two matters is another question and outside my expertise, my access to resources to research and the scope of the present volume.

To cap off the association between films, melancholia and blue, and to come right up to the recent past, in Lars van Trier's 2011 science fiction film *Melancholia,* the second film in his "Depression Trilogy," the eponymous and mysterious rogue planet threatening to collide with planet earth is, of course, blue, as planet earth is predominantly blue too, as Pastoreau (2001, 178–179) points out in the famous NASA photos of "the blue planet." Perhaps van Trier splits off the imaginary rogue blue planet Melancholia from the real predominantly blue planet Earth as a defense for his melancholia about, and mourning for, the Earth.

Black and white moral qualities play out in these films and with the old films being shot in black and white. Similarly, swamps are often color-coded, not only in romantic terms as blue lakes, but also denigrated (literally "blackened") in misogynist, racist and placist terms as bad black waters and contrasted with good blue and white waters (Giblett 2007; 2009, chapter 11; 2013a, chapter 14). The disappearing and ruining of wetlands, such as the Blue Lake, is a black water mass murder, or black mass water murder, or genocide, the killing of a genus of water bodies, what I call "aquaterracide" (Giblett 1996, 2016).

DANCING WITH WATER

The tendency to abstract specific and concrete bodies of water (such as wetlands and rivers) into water as a resource, risk or threat to be managed or exploited is the stock-in-trade of urban planning and landscape architecture if Dovey's and Jones' edited collection is anything to go by, rather than seeing water as the lifeblood of the living body of the earth (Giblett 2007; 2009, 174–189). Similarly, the contributors to *In Time with Water: Design Studies of 3 Australian Cities,* including Melbourne (Bertram and Murphy 2019), abstract specific and concrete bodies of water (wetlands, rivers, etc.) into the overarching category of "water," while also acknowledging "Melbourne's swampy lowlands" in general and the specific wetlands that were on the site of their design case studies and incorporating them into their redesign of their precincts in what could be called "wetlandscape architecture" that not only provides ecological services (such as flood mitigation) and recreational amenities in an aesthetically appealing urban environment, but also echoes the historic swamp and reinstate a remnant of it.

The contributors to *In Time with Water* also color-code water (as we will see below) and neglect Aboriginal stories of the Bunjil, the all-creator, including of the Yarra River, Melbourne's wetlands and the site for the city (Massola 1968, 40, 44, 55–57). The Bunjil is often regarded as an eagle-hawk or as a wedge-tail eagle. As the all-creator, it can assume many forms. It is figured as humanoid in the only visual representation to survive in a cavern or rock shelter in the Black Ranges of western Victoria (figures 7.1 and 7.2). The Bunjil created the underlying structures of land formation that formed the swamps.

Wetlandscape architecture exemplified in *In Time with Water* is a *pas de deux* with the watery lifeblood of the earth body that: dances in time with water in a life-dance with lost and remnant wetlands; refuses the urban choreography of the dance of death on the grave of lost wetlands; acknowledges the materiality and history of specific bodies of water, and; designs urban precincts in time with them and in the context of their catchment or watershed whose topography still largely dictates the flows of water.

For instance, the Arden Morass and Macaulay Common ("Arden Macaulay" for short) in North Melbourne sit at the bottom of the Moonee Ponds Creek catchment that originally flowed into West Melbourne (or Batman's) Swamp (Bertram and Murphy 2019, 124–127). Moonee Ponds Creek at European settlement was a "chain of ponds" (Massy 2020, 130), a seasonally connected

Figure 7.1 Bunjil's Cave or Shelter, Black Range, Victoria. *Source*: Photo by Rod Giblett

Figure 7.2 Interpretive signage for Bunjil's Shelter, Black Range, Victoria. *Source*: Photo by Rod Giblett

and disconnected series of wetlands remarked upon by early European explorers elsewhere in Australia.[1] The Aboriginal name of "Moonee Moonee" indicates as much. They were "channelised" and connected to the "canalised" Yarra River, while West Melbourne (or Batman's) Swamp was drained and filled in a "cut and fill" civil engineering operation that involved cutting down Batman's Hill and filling up Batman's Swamp (Giblett 2016, 2020c). Additional challenges for the site are the presence of the Upfield railway line and the CityLink tollway, both of which follow the course of the Moonee Ponds Creek for some way (Bertram and Murphy 2019, 125). They also divide both sides of the creek from each other.

The starting point for the Arden Macaulay design investigation was the desire to offer "a more transformative and projective vision" than "the numerous policies, plans, and strategies from government and industry for this brownfield site that are currently available by approaching the redevelopment of Arden Macaulay in relation to its particular physical characteristics and 'personality'—its topography, its geology, its infrastructure, its water bodies, its industrial heritage" (Bertram and Murphy 2019, 124). In other words, approaching, unlike *Urban Choreography*, the redevelopment of Arden Macaulay in relation to both its bioregional ecology and its environmental history as a living entity with agency. The bioregional ecology of the site includes its present situation in a catchment with its distinctive hydrogeology and geomorphology. The environmental history of the site includes its past ecology within the context of colonization, canalization, and drainage or filling of Melbourne's wetlands, rivers, and creeks.

The Arden Macaulay design investigation ran over 18 months with "a mixed method" approach of assembling archival and contemporary sources, including old maps and documents and whizz bang, new GIS modeling. Out of this process, the Arden Macaulay design investigation proposes the transformation of the creek/drain into artificial wetlands. These comprise what Bertram and Murphy (2019, 125) describe as "three distinct ecological vegetation class zones" produced by "the introduction of localized dams [thus] enabling the reinstatement of wetland ecologies and physical connectivity across the divided Moonee Ponds Creek." The result would be what they go on to call "islands of intensity" with "vegetated wetlands" "echoing the historic swamp and chain of ponds" thus reinstating a remnant of West Melbourne Swamp (Bertram and Murphy 2019, 125–126).

In listening to these artificial wetlands, the listener might hear an echo of the historic swamp and chain of ponds that were there once upon a time if they have their ears attuned to hear the echo. The construction of what they call "Swamp Pavilions" would "provide immersive experiences of the swampland" (Bertram and Murphy 2019, 126). They could retell Aboriginal stories of the Bunjil. They could also include interpretation of Melbourne as a city of wetlands past and present and for the future. They might enable the listener to hear the echo of the historic swamp and chain of ponds in what Walter Benjamin (1999, 462, 464) called "a dialectical image" in which "the what has been" would be echoing with "the now of recognisability." The "Swamp Pavilions" could also include archival and contemporary maps of the historic swamp and chain of ponds in "a dialectical image" of the area in the past with its wetlands and in the present without its wetlands.

A dialectical image would present the absence and what has been of Melbourne as an Aboriginal estate of swamp and marsh created by the Bunjil, and in the presence and now of the recognisability of Melbourne as city and metropolis from which Aboriginal owners and wetlands are absent, coming together, or "synthesised," as marsh metropolis and swamp city, what I call "aquaterrapolis" (Giblett 2016, chapter 2). Such a dialectical image might enable the listener and viewer to hear, see, and appreciate the ecology of the bioregion in which they live and its environmental history so that they might live bio- and psycho-symbiotic livelihoods in their bioregional home habitat of the living earth (Giblett 2011, chapter 12).

Such an appreciation and way of living would entail a reconfiguration of ecology and economics as they are usually construed (Giblett 2011, chapter 2). The terms "ecology" and "economics" come from the Greek "*oikis*," usually translated as "house" or "household," but translated by Benveniste (2016, 253) and his translator into English as "habitat." On this definition, ecology is the study of habitat and economics is the management of habitat. Or more elaborately, ecology is the study, and economics of the management, of plants and animals in their habitat, and as the habitat for other plants and animals (including humans). Habitat is the bioregion and home is the bioregion (Giblett 2011, chapter 12).

A much warmer, less scientific, albeit archaic term that combines all these features into a single term is what Hills (1975, 1) calls "abode," so that the earth on, beneath and above its surface is considered as what he goes on to call "the abode of living organisms." The lithosphere (rock-sphere), the hydrosphere (water-sphere), pedosphere (soil-sphere), and the atmosphere (air-sphere) go together to make up the biosphere (life-sphere) and the ecosphere (habitat-sphere), the earthly abode.

Such an appreciation would also entail a reconfiguration of the city. The resulting city arising out of the design investigation for Arden Macaulay would be what Bertram and Murphy (2019, 127) describe in conclusion as "a new kind of contemporary city with a unique identity of water [and water bodies, such as wetlands lost and found] at its core" designed and constructed in wetlandscape architecture and dancing a *pas de deux* with water as the lifeblood of the earth body. Other cities, such as Auckland with its North Wharf, Shanghai with Houtan Park and San Francisco with Crissy Field or Park, have successfully developed and applied similar principles of wetlandscape architecture in designing and constructing artificial wetlands in abandoned wastelands (Thompson 2014, 92; Waterman 2015, 70–71, 88–89; Wraight and Associates, n.d; Hargreaves Associates, n.d.; Golden Gate National Parks Conservancy, n.d.; Landscape Architecture Foundation, n.d.).

A reconfiguration of the city could also entail how the city is seen and portrayed in a display of archival maps of Melbourne in a "Swamp Pavilion." Bertram and Murphy (2019, 106–107) construct a dialectical image of Melbourne based on the 1866 British Admiralty map of Melbourne (Cox 1866) showing West and South Melbourne Swamps (and many other swamps) and the old course of the Yarra River (shown in light blue in their reproduction). These are overprinted in dark blue in their reproduction showing the reconfiguration of various water bodies in the "subsequent engineered modifications carried out in the 1880s and 1990s," such as the straightening and rerouting of the river and the construction of docks (with water shown in red in the new docks in the original of a later map of 1879; Coode 1879). The light blue in Bertram's and Murphy's map shows the course of the river that has been lost, while the dark blue shows the river and docks that were gained. The use of blue is a reprise of the gendered color-coding of waters with the dialectic image of a light, romantic wet daydream of bygone waters and a dark, modern nightmare of waters lost and ruined in wastelands.

The light blue also shows the swamps in 1864, most of which have been lost, including the "Blue Lake" of West Melbourne (or Batman's) Swamp. This, and other lost wetlands, such as Sandridge Flats, later Fishermens Bend (Giblett 2020c, chapter 4), could have been depicted in another color, such as rusty red to signify that these wetlands were ruined as waste dumps and became waste lands (as McCrae indicated with the "Blue Lake"). The exception among the swamps mapped in 1864 (Cox 1866) that weren't lost and can still be found today is South Melbourne Swamp, albeit landscaped into, and reincarnated as, Albert Park Lake. This swamp/lake should have been depicted in the darker blue to indicate that it, too, was subjected to engineered modifications, albeit in the 1930s (Giblett 2020c, 30). This dialectical image would also be associated with the seductive, but dangerous, femme fatale of the blue angel of the "Blue Lake."

Dean Stewart (2020) also depicts the largely lost wetlands of Melbourne in blue in his map of Aboriginal Melbourne, also based on the 1866 British Admiralty map (Cox 1866), like Bertram's and Murphy's rejigged map. This map is an idealised and idyllic (even pastoral) portrayal of Melbourne, its wetlands and grasslands, depicted along the lines of the colored version of J. R. R. Tolkien's map of Middle Earth (Dean Stewart, pers. comm.). Stewart's (2020) map portrays Melbourne in nostalgic, bucolic, idyllic terms, just as Tolkien's Middle Earth phantasies England in the Middle Ages. Both have their dark places too. Tolkien's Middle Earth included the Dead Marshes, a war-torn wetland based on his own experience of the wet wasteland of the Western Front in the industrial warfare of World War I (as we saw in chapter 1). Similarly, Melbourne eventually included the wet wasteland of Fishermens Bend in the industrial war against wetlands (Giblett 2020, chapter 4). Stewart's (2020) map of Melbourne shows Aboriginal figures who owned these wetlands and drylands and intimates that genocide took place alongside aquaterracide. Aboriginal people, like the wetlands, are largely absent from this place. This map also shows the "Hoddle grid" of the CBD devised in 1837, two years after the start of settlement. It does not show the extent of urban development as shown in the 1866 British Admiralty map (Cox 1866) on which Stewart (2020) based his map. Stewart freezes a moment in the past, as all maps do, in his case, thirty years before the time of the British Admiralty map (Cox 1866).

Depicting water bodies, such as rivers and wetlands, in blue on maps is a convention of cartography that began in the late fifteenth century. Pastoreau (2001, 181) argues that "on the oldest maps [. . .] water [. . .] was almost always green." On the 1879 map (Coode 1879) and on the newest maps, as we have seen, water is usually depicted in blue. It is a product of what Pastoreau (2001, 181) calls "the common association of blue with water." He concludes that "it took quite a while for water to become blue, and for blue to become cold" (Pastoreau 2001, 181). The shallow waters of wetlands can be hot though. Depicting them in blue (and so as cold) on maps is misleading, though it is appropriate for the hot/cold femme fatale in black and white film. It will take quite a while for wetlands to cease to be aestheticized as blue waters or denigrated as black waters and to be celebrated as living black waters (Giblett 1996).

THE WATERY FABRIC OF SPACE

Along similar lines to Dovey and Jones in *Urban Choreography*, Matthew Gandy's 2014 book of urban geography, *The Fabric of Space*, generally abstracts specific, concrete bodies and places of water into the spaces of water wherein water is a resource and commodity, while also intermittently presenting some specific case studies of wetlandscape architecture along similar lines to Bertram and Murphy. Water in *The Fabric of Space* is implicitly the fabric of space, but nowhere in this book is this figuring of water as fabric in its title defined or discussed. Perhaps the fabric of space of the modern city that holds it together for Gandy is more precisely the infrastructure of the warp and woof of steel water pipes (as in Mumbai as depicted on the cover of *The Fabric of Space*), concreted, canalised rivers (as in Los Angeles River and as depicted in photos and film stills) and brick-lined underground sewers (as depicted in Nadar's photos of them in Paris).

The Fabric of Space considers the supply of fresh water in pipes, the disposal of wastewater in sewers, and the display and transport of water in rivers in case studies of Paris, Berlin, Lagos, Los Angeles, Mumbai, and London. Each account of these cities begins at some arbitrary point in the past of high modernity moves forward into the present and speculates about the future. The beginnings of Paris, Berlin and London in the premodern wetlands on, or in, which these cities were founded, are (in the case of Paris and Berlin) ignored or overlooked, or (in the case of London) mentioned occasionally (for Paris, London and Berlin as wetland cities, see Giblett 2016, chapters 3, 4, and 6). The swamps of Lagos in Nigeria are still visible today so they cannot be overlooked or ignored, whereas the wetlands of Paris, Berlin, and London are invisible (lost, absent, drained, filled, and destroyed) so they can be overlooked or ignored along similar lines to Melbourne in *Urban Choreography*. *The Fabric of Space* is concerned with the visible or invisible present fabric of the spaces of water, and not with the invisible and lost or absent places of water.

The lost wetlands of London on which it was founded in the past have left a trace in the present and in envisioning the future; *The Fabric of Space* acknowledges both. The final chapter on London begins by referring to J. G.

Ballard's 1962 dystopian science fiction, *The Drowned World,* and Richard Jefferies' 1885 pioneering post-apocalyptic fiction, *After London.* Both novels are set in the future in which the city is, as Gandy (2014, 186) puts it, "reclaimed by the swamps it was built on"—by reclaiming them. The city was built by reclaiming the swamps (Giblett 2016, chapter 4). In both novels, the swamps re-reclaim the city. Along similar lines, George Turner's prescient *The Sea and Summer* published in 1987 envisions Melbourne in the future when the emission of greenhouse gases in the age of climate change and global warning has caused the polar ice sheets to melt, sea levels to rise and the city to flood (Giblett 2020c, 137). Bertram and Murphy (2019, 127) acknowledge this possibility for Arden Macaulay, but not Dovey and Jones for Melbourne.

The Fabric of Space traces how the development of London involved "human modifications of the riverscape," and wet landscape, including "extensive [wet]land reclamation from the salt marshes within the river's floodplain toward the east" (Gandy 2014, 188). The Thames River today is "effectively a tidal canal" whose "tidal floodplain now comprises some 35,000 hectares of low-lying land" (Gandy 2014, 189). "Low-lying land" is a pejorative term for former (drained or filled) wetlands in a hierarchy of moralized land values (low is bad, high is good). In the age of global heating and climate catastrophe the danger is that "the Thames would begin to revert to its much and more unpredictable course with a greatly expanded floodplain" (Gandy 2014, 187). In other words, revert to swamp and marsh. The Thames Barrier, as a monument of techno-scientific engineering design in a visible, aesthetically pleasing and sublime object, is designed to prevent the uncanny inundation of the city by flooding in an aesthetically displeasing and slimy abject in a return of and to its repressed wetlands.

Remnants of the Thames wetlands remain in its estuary. Rainham Marshes was a military site and is now a nature reserve. *The Fabric of Space* relates how "the conversion" of these Marshes from sinful site into saved nature "represents only a small fragment of the riverside landscape" (Gandy 2014, 203). Or more precisely, it represents only a small fragment of the riverine wet landscape in which there is now here, and was elsewhere, no hard and fast dividing line between river and floodplain, between water in a river and in a wetland. Water does not know or respect the difference between river and wetland. A river is not merely a conduit for transporting water and vessels, nor an alimentary canal for conveying life-giving nutrients and death-dealing wastes, but a body of water embracing moving flows, stagnant pools and plains above and below ground.[2]

The Fabric of Space goes on to give a valuable, extensive and critical account of the "ecological restoration" (Gandy's scare quotes) of these marshes in "landscape design" (Gandy 2014, 207, 203–208 for the account), or more precisely "wet landscape design" as landscape and wetlands have been inimical to each other (see Giblett 1996, 2011), and as wetlands pose specific challenges and opportunities for design compared to drylands (as we have seen in this chapter and in the previous chapter). *The Fabric of Space* dances around the specificity of the geography and history of London as a wetland city in an urban choreography (along similar lines to Dovey and Jones in *Urban Choreography*), rather than dancing with the wetland city and with water as the lifeblood of the earth in a *pas de deux* (as Bertram and Murphy do in *In Time with Water*).

The "Epilogue" to *The Fabric of Space* is devoted to a discussion of "landscape design" in "new waterscapes" of "urban wetlands" in London and other cities, such as Singapore (Gandy 2014, 218). *In Time with Water* provides examples of Australian cities, such as Melbourne, Perth and Brisbane. Other examples could have included North Wharf in Auckland, Houtan Park in Shanghai and Crissy Field or Park in San Francisco. Crissy Field or Park was, like Rainham Marshes, a military site that had been salt marsh and estuary, and was re-reclaimed as an urban wetland. By producing a chronological history of water and some cities beginning at an arbitrary point in the past, proceeding to the present, and projecting into the future, *The Fabric of Space* concludes where it might have begun with wetlands and wetlandscape architecture that would present a genealogical "history of the present," as Michel Foucault (1977, 31) calls it, of water and cities that acknowledges its case studies cities as wetland cities with flows and bodies of water in the past, present and for the future, as *In Time with Water* does with Melbourne and Perth. A genealogical "history of the present" of water and cities, of wetlands past and present, lost and found, absent and present, would also appreciate and value the ecology and economy of the bioregional home habitat and water as the lifeblood of the living earth. Long live living waters!

NOTES

1. As we have seen in chapter 4. For Moonee Ponds Creek as a chain of ponds and called "Moonee Moonee Chain of Ponds," see Crofs' map of 1838 (Bertram and Murphy 2019, 140).

2. Compare a similar situation with the Fraser River in British Columbia in Canada and a critique of an account of its geography and history that ignores or overlooks its wetlands, including in its estuary (Giblett 2014, esp. 117), and contrast that with other rivers in Canada, such as the Don and St Lawrence, and the accounts of their geography and history that do not ignore or overlook their wetlands, such as Ashbridge's Bay Marsh and the Kamouraska Marshes (Giblett 2014). Compare also the Swan River in Perth, Western Australia, which early European explorers saw as only an alimentary canal for transporting nutrients and wastes down a digestive tract that ignores the supporting wetlands of the kidneys and liver of the body of the earth (Giblett 2008, chapter 11; 2013a, chapter 15).

Part IV

WETLANDS AND NATURE WRITING

Chapter 8

Henry David Thoreau

The Patron Saint of Swamps

Thoreau is famous as a nature writer and environmental philosopher of the wilds and the woods, an early uptaker and proponent of evolutionary biology and a pioneer of forest ecology. He has also been dubbed the "patron saint of swamps" (Giblett 1996, ix, 229; 2020c, chapter 4) as he reflected and wrote philosophically on bogs, marshes and swamps. though most writers about Thoreau, including all his biographers, neglect or overlook this aspect of his writings. Two or three of the too few Thoreauvians *not* to do so are Branka Arsić (2016, 222–225), Carl Bode, and Andrew Menard. Bode (1982, 686) notes Thoreau's "love for swamps" and how "he enjoys being in them, enjoys writing about them." "Swamp" for Menard (2018, 176) is "a defining figure in Thoreau's work" according to the index of *Learning from Thoreau* without elaborating much in the body of the book on what this means. The index entries refer to pages in which Menard (and Thoreau) mention swamps, and not to a discussion of the swamp as "a defining figure."

The present chapter partly aims to remedy the neglect or oversight of this aspect of his work. It also partly aims, not only to supply this lack or to fill this absence, but also to demonstrate the importance of Thoreau's writings for critiquing the conventional cultural construction and perception of wetlands in mainstream western culture. Marshes and swamps were a persistent topic before and during his two-year, two-month, and two-day sojourn at Walden Pond. They were the womb and birthplace of American conservation, whereas Walden Woods and Pond were its later cradle after it had been born and where it was nurtured. When he also wrote later and lovingly about swamps, such as in "Walking," he upset and overturned mainstream thinking about swamps, including the conventions of European landscape aesthetics, philosophy, medicine, and psychology that regarded them as dismal, disease-ridden, and displeasing. In this essay and other writings, Thoreau developed a conservation counter-aesthetic of swamps and a counter-Christian religion and theology of swamps. He valued swamps as quaking zones where body and mind meet in the premodern wetland. As a proponent of evolutionary biology and as a non-believer in the institutionalized religion of Christianity, he developed his conservation counter-theology against creationism and for evolution in swamps.

Walden, his masterpiece, is largely to blame for this neglect or oversight regarding the importance of wetlands for Thoreau, especially as the first edition bore the subtitle, *or Life in the Woods.* He never wrote a book with the title and subtitle, *Concord, or Life in the Marshes and Swamps*. It is up to others to try to follow in his footsteps into the swamp and wade into marshes as he enjoined in *Walden* (Thoreau 1982, 557). Thoreau was a writer about wildness first and foremost, irrespective of whether wildness was in the dark woods, or in the dismal swamps and monstrous marshes, or on a sublime mountain, or in "the bog in our brains and our bowels," as he puts it (Thoreau 1962, *IX*, 43). Thoreau is the epitome of the boggy counter-tradition to the mainstream western cultural constructions of wetlands.

When Thoreau lived at Walden Pond, he also lived close to Wyman Meadow, what is today described as a wetland and is signposted as such on the trail to the site of his house. Yet, while he was living at Walden Pond, he did not write about much about Wyman Meadow. He certainly did not acknowledge in writing the biological richness of this wetland. Although he overlooked this wetland both literally (as he could not see it from his house) and figuratively (as he did not write about it in any detail), this proximity to a wetland may have begun to instill and nurture in him his love for swamps that was developed and distilled in his later writings, especially in his posthumously published essay "Walking" in which the swamp is a sacred place. In this essay and other writings, Thoreau developed a counter-Judeo-Christian religion of swamps as a sacred space.

WYMAN MEADOW, HENRY'S HOUSE AND HIS DESK

On "Patriot's Day," Monday April 15, 2019, I completed the "Thoreauvian pilgrimage" (Buell 1995, 311–338) and walked along the trail by Walden Pond to Henry's original house site and then finally on a boardwalk over a swale to Wyman Meadow that connects the meadow to Walden Pond. The signage located there says that this meadow is a wetland. Indeed, it was literally wetland when I was there with water on the surface. Thorson's (2018, 104) photograph of Wyman Meadow shows it dry, whereas it was wet when I was there in April 2019. "Wetland" is, of course, a twentieth-century term and catch-all category that embraces variously what Thoreau and others in his time and before called bog, swamp, marsh and meadow, with some slippage in meadow as cultivated (i.e., field) or uncultivated (i.e., marsh) and between marsh and meadow (as we will see with Thoreau; see also Hampl 2006, 223–224).

I had an epiphany when I was there in that place that "the patron saint of swamps" (Giblett 1996, ix and 229–239) lived close to the wetland of Wyman Meadow during his sojourn at Walden pond—a fitting locale for such a swampy saint. This perhaps fortuitous or unconscious locating of his house near a wetland manifested Thoreau's love of swamps and other wetlands, such as meadows, that is absent from *Walden*, but is present in his other earlier and later writings. In "Walking," for instance, Thoreau (1982, 612) encourages us to "bring your sills up to the very edge of the swamp, then (though it may not be the best place for a dry cellar)." Which is exactly where Thoreau located his house on a rise next to the wetland of Wyman Meadow, but not so close or low that he did not build his house with a dry cellar too, nor so high that he could see the meadow from his house.

Robert Thorson, the author of *The Guide to Walden Pond* and other exemplary books about Thoreau, confirmed the closeness of his house to the wetland of Wyman Meadow: "From Thoreau's door, he could take about 10 steps east to the top of the small rise and see the whole thing. The edge of Wyman meadow was only half as far as the edge of Walden Pond at the Waterfront, where Thoreau usually went."[1] In other words, he did not usually go to Wyman Meadow. Perhaps he did not need to go there because he was already there, or close to it anyway. While he was living by Walden Pond, Thoreau usually wrote about Walden Pond and Woods, and not about Wyman Meadow. Thoreau does not explicitly refer to Wyman Meadow by name in *Walden*. He mentions Wyman the potter and squatter after whom the Meadow is named and he refers to the meadow in passing elsewhere as a marsh. For instance, he describes in *Walden* how "a mink steals out of the marsh before my door and seizes a frog by the shore" (Thoreau 1982, 366). The marsh was in fact not before his door, but before the window on the bed (or east) side of his house (as we will see later in the present chapter).

Elsewhere in *Walden* Thoreau calls this marsh a meadow. He describes how "the marsh-hawk sailing low over the meadow is already seeking the first slimy life that awakes" (Thoreau 1982, 550). He does not elaborate, let alone celebrate, the slimy life of the meadow as he would do later of various swamps, real and imaginary. Thorson (2018, 109) in his *Guide* connects Thoreau's comments in *Walden* about the pond's croaking frogs with the fact that Wyman's Meadow would have been an important breeding ground for frogs. Their croaking at Wyman's Meadow would have been audible from Thoreau's doorstep and could have inspired those passages. Thoreau overheard the frogs as he overlooked the meadow—perhaps a case of not being able to see the marsh for the sound of the frogs? Thoreau certainly celebrated in his later writings the slimy life of marshes and swamps.

Of Wyman Meadow, Thoreau describes how the "place was long since converted into a meadow" (Thoreau 1982, 429). Thoreau does not say from what the meadow was converted. His previous mention of the mink (and even the marsh-hawk) suggests the place was converted from a marsh into a meadow. Thoreau (1982, 276) wrote earlier of Walden Pond that there were "no Neva marshes to be filled" suggesting that the place had already been filled to create Wyman Meadow by the time of his sojourn there. He does not discuss, let alone bemoan the fact, that the marsh was converted into a meadow, if it were a marsh in the first place, or explain how this conversion took place, nor mourn the death of the marsh. This conversion most likely involved clearing the native vegetation to expose the clay that Wyman and his son dug from "the pond's silty banks" (Walls 2017, 199) and used in making their pottery. The swale that now connects the meadow and the pond was perhaps dug out of a mistaken wish to drain the marsh into the pond with a lack of knowledge about hydrogeology. The marsh was not, and probably could not be, drained artificially and dried out completely on the surface as the marsh and pond were the expression of the same, contiguous body of water, or aquifer, at the same level superficial, visible with an underground flow of water from meadow to pond, north to south. The waters of Walden Pond and neighboring Wyman Meadow are not only contiguous, but also continuous, one body of water.

For Thorson (2018, 104 and 105) in his *Guide* Wyman Meadow is "by far the ecologically richest part [. . .] of Walden's Shore." This is not surprising as it is a wetland. Wetlands are among the most richly biodiverse of all ecosystems on the planet earth, if not the most biodiverse.[2] Thorson (2018, 110) later contrasts "the biological richness of Wyman Meadow" with "the limited biodiversity near the pond path" leading to Thoreau's house site on the north side of Walden Pond starting from the Visitors' Center, Thoreau Society shop and car park. Thoreau was aware later that swamps and meadows are biologically rich (and probably ahead of his time scientifically in this regard), but while he was living and writing by Walden Pond from 1845 to 1847 he did not comment on the fact that the meadow almost literally on his doorstep was the most biologically richest part of Walden Pond and Woods.

Thoreau mentions the view of the Pond in *Walden*, but not the view of Wyman Meadow. Thoreau (1982, 424) wrote in *Walden* that the scenery of Walden (meaning the Pond and its surrounding Woods) is "on a humble scale, and, though very beautiful, does not approach to grandeur." Thoreau is looking at the visible surface of the land and water, at landscape, an object of aesthetic appreciation with him as its subject. Subject and object are set up in their mutually constitutive positions. He is not considering the hidden depths of the wetland as an abject (to use Julia Kristeva's term (1982)), or as an agent of ecological processes with himself as abject too. The scenery of the wetland of Wyman's Meadow for Thoreau is obscene in the sense of being hidden and secret, out of sight and out of mind. The obscenery and abjectness of marshes and swamps only come out into the open in Thoreau's later writings where it upsets the conventions of the European landscape aesthetic and evokes participation in ecological processes.

While I was visiting Walden Pond, I started to wonder if Thoreau's desk may have been placed under the window on the eastern side of his house (as it is in the replica near the Visitors' Center) facing the wetland and that he may have been able to view the wetland from this window as he wrote at his desk.[3] I also asked Robert Thorson about this and he kindly corrected me: "In the house replica, they have the desk facing east, on the starboard side of the house facing inward. In *Walden* Thoreau makes it clear that his desk is on the port (left, west) side when it was on site, facing a woodpile and clearing. The other (east) side is the edge of the hollow, an earthen bank rising to block any view."[4] Indeed, in *Walden* Thoreau (1982, 496) writes that "every man looks at his wood-pile with a kind of affection. I love to have mine before my window." His house had two windows with his woodpile outside the window under which his desk was located where he wrote this, while his bed was on the other, Wyman Meadow side.

During his sojourn at Walden Pond in Walden Woods Thoreau did not write that he loved having a meadow (or wetland) before his other window, nor did he look at it with a kind of affection. Thoreau overlooked (in a variety of senses) the wetland of Wyman Meadow: the view from his house looked over it as it was too low to see it down in its hollow; in *Walden* he overlooked the biodiversity of the meadow as a wetland as he did not comment on it. Thoreau's love of marshes and swamps and looking at them with affection had been expressed earlier in his *Journal* and was later in this and his other writings (as I will show below in the present chapter).[5]

It was only by being there in the place, sauntering by the pond, walking the trail to the house site, stopping and seeing a sign that said Wyman's Meadow is a wetland that I had this epiphany about the proximity of Henry's house at Walden Pond to a wetland. This perhaps fortuitous locating of his house by a wetland may have worked on him unconsciously when and where he lived by Walden Pond from 1845 to 1847. It blossomed into a conscious love of swamps in his later life and writings, such as in his *Journal* and in his late and posthumously published essay "Walking."

THE SWAMP IS THE HOLY OF HOLIES

Thoreau not only wrote on swamps, but also maintained that swamps wrote on him, or at least on his heart. He said that "when I will die, you will find swamp oak written on my heart" (cited by Richardson 1993, 17). The swamp not only wrote on Thoreau; he also wrote repeatedly on it. Thoreau (1962, *II*, 13) wrote in his *Journal* in 1850 that "I have been surprised to discover the amount and various kinds of life which a single shallow swamp will sustain." The same could be said about Wyman Meadow. Thoreau goes on to discuss a meadow on "the south side of the pond" (and not Wyman Meadow on the north side). Two years later Thoreau was no longer surprised by the biological richness of the swamp. Thoreau (1962, *IV*, 281) wrote in his *Journal* on August 4, 1852, that the swamps are "the wildest and richest gardens that we have. Such a depth of verdure into which you sink." Thoreau was no mere walker by the wetland, but a wanderer in the wetland who was not afraid of sinking into it, provided he eventually found "a hard bottom" somewhere (as we will see later). It took another five years for Thoreau (1962, *X*, 196–197) to describe in his *Journal* the plants of a swamp as he did with Gowing's Swamp on November 23, 1857.

The richness and verdure of the swamp made it into his "real garden." For Thoreau (1993, 198) writing in 1860–1861 "some rich withdrawn and untrodden swamp [. . .] is your real garden." In "Walking," his posthumously published last essay containing his most memorable pronouncement on swamps, Thoreau (1982, 612) stated that "I derive more of my subsistence from the swamps which surround my native town than from the cultivated gardens in the village." The swamp for Thoreau became a place that gave both physical and spiritual sustenance. On January 4, 1853, Thoreau (1962, *IV*, 449) declared in his *Journal*"my temple is the swamp." In "Walking" Thoreau (1982, 613) describes how, "[w]hen I would recreate myself, I seek the darkest wood, the thickest and most impenetrable and to the citizen, most dismal, swamp. I enter a swamp as a sacred place, a *sanctum sanctorum*." Thoreau is not merely alluding to, and countering, the Great Dismal Swamp on the border between Virginia and North Carolina, but refusing and inverting (if not subverting) the solid citizen's, and the dominant cultural paradigm's, view that all swamps are dismal by regarding them as the most sacred of places, as the holy of holies as viewed by the alternative cultural paradigm and the slimy denizens of the swamp (see Giblett 2011, chapter 1 for the two cultural paradigms).

Thoreau (1982, 613) sought refuge and renewal in the swamp as a sacred place, indeed as the inner sanctum, the holy of holies into which, like the Jewish High Priest, he would "annually go on a pilgrimage" (Thoreau 1993, 197). He would perform the ritual of life-giving, self-baptism in the swamp whose waters were not rank poison: "far from being poisoned in the strong water of the swamp, it is a sort of baptism for which I had waited" (Thoreau 1962, *IX*, 376–377). Thoreau was a swamp self-Baptist. As an ex-Baptist myself, a swamp self-Baptist is the kind of Baptist I now want to be.

In the swamp, rather than at the pond, he found religion because this is where he found love. Thoreau (1958, 52; 2013, 89) wrote in a letter dated

September 8, 1841, that "our religion is where our love is." God is not love (*contra* the New Testament Epistle of 1 *John* 4: 8 in which "God is love"), but religion is love, or more precisely where our love is for Thoreau. Rather than being merely contrarian (as he was sometimes accused of being), for him this meant being and living affirmatively at home in one's body, time and place.[6] Such places for him were Walden Pond and Concord swamps. When one finds love, one finds religion.[7] Thoreau found love in a swamp, but also faith and hope. Thoreau (1993, vii) famously had, professed and proclaimed faith in a seed, but he also had faith and hope in a swamp. Faith, hope and love are the three primary fruits of the spirit according to St Paul (1 Corinthians 13:13); for Thoreau, they are the fruits of the swamp. Swamp is spirit in Thoreau's theology. As spirit is for St Paul, swamp is for Thoreau. Thoreau's preference for swamps in "Walking" over town gardens was no mere nostalgia for a lost pastoral paradise as "hope and the future for me are not in lawns and cultivated fields, not in towns and cities, but in the impervious and quaking swamps" (Thoreau 1982, 611), such as in the wetland of Wyman Meadow, other marshes and living black waters.

Despair is black (as in Shakespeare's *Richard III* discussed in chapter 2), but black is not despair. Rather, black is hope. Hope lives in the depths of living black water; the heights of hope spring eternal from the depths of despair in the slough of despond. The eternal now irrupts in chronological time. Thoreau deconstructs the metaphysical hierarchy of values of high over low, upper over lower, white over black, light over darkness, present over past, present over absent. He helps with the construction of the central argument (in this book) against the mainstream western denigration of wetlands and makes the case compelling for their conservation. This case is cultural—agricultural, aquacultural, horticultural—in a word, ecocultural, the culture of the earthly household. The depths of hope in the slimy black waters of the swamp undermine the heights of despair in the sublime city and in the dead black waters of its slums and industrial wet wastelands. Hope springs eternal from the grotesque lower bodily and earthly strata and subverts the heady heights of sublimated theory.

For Thoreau (1982, 613), the swamp is "the strength, the marrow of Nature." The strength of nature, for him, lies not in the hard bones of the dry land, but in the soft marrow of the marshes and swamps, what he also called the liquor of nature that feeds the earthly body, the body environmental: "the very sight of this half-stagnant pond-hole, drying up and leaving bare mud [. . .] is agreeable and encouraging to behold, as if it contained the seeds of life, the liquor rather, boiled down. The foulest water will bubble purely. They speak to our blood, even these stagnant, slimy pools" (Thoreau 1962, *IV*, 102). Unlike the stagnant pools of city slums with death-dealing water that repulsed Engels, the stagnant pools of country swamps for Thoreau speak to our blood because they contain live-giving water. Water for Thoreau (1962, *XIII*, 163) is "the most living part of nature. This is the blood of the earth." Water for Thoreau is the life-blood of the body of the earth. Thoreau sees the earth as body as traditional, premodern cultures do (see Giblett 2008a).

Thoreau's blood circulates with the blood of the earth and with the liquor and marrow of swamps in the body of the Earth: "surely one may as profitably be soaked in the juices of a swamp for one day as pick his way dryshod over sand" (Thoreau 1982, 187). Thoreau would prefer the problems of travel through the wetland, the marrow of the earth, to the ease of passage over the dry land, the bones of the earth. Yet the problems of travel across the wetland are seasonal anyway in the higher latitudes as "the deep, impenetrable marsh, where the heron waded and bittern squatted [in summer], is made pervious [in winter] to our swift shoes, as if a thousand railroads had been made into it" (Thoreau 1982, 71). Thoreau sees himself as part and parcel of nature, as circulating in the body of nature, not only via the circulatory system of rivers (as in his journey on the Concord and Merrimack Rivers), but also in the semi-stagnant system of marrow through immersion in the swamp by a kind of secular baptism.

Without the wetland, without wetlands, the world would fall apart. The wetland feeds and holds together, nurtures and coheres, the skeleton of the body of nature. Without the wetland, there would be nothing to replenish the skeletal system of the dry land, the backbones of mountain ranges, the ribs of ridges, the limbs of peninsulas and capes, and the fingers of land reaching into the sea, all of which (including the marrow of the wetlands) supply and make possible the fertile plains, prairies, and steppes on which agriculture takes place, on which the dominant cultural paradigm relies, on which industry depends, on which cities "live," or more precisely which they parasitically suck dry.

Thoreau prefers the resources of hope cultivated by the alternative cultural paradigm in the native swampy quaking zone in the country to the monuments to the past enshrined by the dominant cultural paradigm in the horticultural, agricultural, architectural, industrial, and feral quaking zones in the city. Thoreau's rhetorical tactic against the stratagems of standard swamp-speak within the dominant cultural paradigm was to displace and upset the usual or normative disjunction between swamp and garden by seeing the swamp as garden, and so exploit the favorable associations of the garden as a place of light and life (see Giblett 1996 for standard swamp-speak and Giblett 2011, chapter 1 for the dominant [and alternative)] cultural paradigms).

Thoreau's writings on swamps touch on and counter, or subvert, many aspects of standard swamp-speak in the dominant cultural paradigm of mainstream western culture that views marshes and swamps in denigratory terms as places of darkness, disease, horror and the monstrous—the obscenery of marshes and swamps, or "nature's shames and ills" as Charles Cotton (cited by Giblett 2011, 65) called them—that should remain hidden and not come into the open, but which do come out in Thoreau's writings on swamps and in the present chapter. Although Thoreau by no means addressed, and countered, every point of standard swamp-speak (as he seemed to be more concerned to produce affirmative press for swamps and support the alternative cultural paradigm, rather than simply rebuff, or even rebut, the denigratory), he did turn the rhetoric against itself.

Thoreau countered the prevailing miasmatic theory of disease (which maintained that one could catch diseases, such as malaria [literally "bad air"], by breathing bad air), by stating how "miasma and infection are from within, not without" (Thoreau 1980, 261; 1962, *V*, 394). He upset this theory by suggesting how "the steam which rises from swamps and pools is as dear and domestic as that of our own kettle" (Thoreau 1982, 61). For Thoreau, swamps and stagnant pools were not the antithesis of, nor a threat to, the homely, but of comparable value. He did not valorize the wetland over the homely but gave them equal value, unlike those of his (and my) contemporaries who denigrated and feared the wetland (and accordingly valorized the canny over the uncanny).

Rather than seeing the airs of swamps as bearers of disease, Thoreau made a crucial distinction between mist and miasma, and even saw mist as healing in a poem that posthumously acquired the title of "Mist" or "Fog":

Low-anchored cloud,
Newfoundland air,
Fountain-head and source of rivers,
Dew-cloth, dream-drapery,
And napkin spread by fays;
Drifting meadow of the air,
Where bloom the daisied banks and violets,
And in whose fenny labyrinth
The bittern booms and heron wades;
Spirit of lakes and seas and rivers,
Bear only perfumes and the scent
Of healing herbs to just men's fields! (Thoreau 1982, 165–166; 2010, 56, 406)

In a poem devoted exclusively to the subject and entitled simply "Fog," Thoreau (1982, 237–238; 2010, 150) referred to it as "dull water spirit—and Protean God," as "incense of earth," "spirit of lakes and rivers" and as "night thoughts of earth." Rather than a vector for disease and a cause of death like miasma, fog is a source of new life. Rather than regarding fog and mist as vapors bearing disease and death, why not see them as the visible manifestation of the exhalations of the earth, particularly of the trees, on which we are dependent for life? After all, we are in symbiosis with the oxygen-producing plants of the Earth.

The swamp vapors were as homely for Thoreau as kettle steam because the swamp itself was better than a homely garden. Indeed, if Thoreau (1982, 612–613) had to choose between them he would have chosen the swamp every time: "yes, though you may think me perverse, if it were proposed to me to dwell in the neighborhood of the most beautiful garden that ever human art contrived, or else of a Dismal Swamp, I should certainly decide for the swamp."

Rather than a reason for avoiding the black swamp of despair and depression and the blue swamp of melancholia, Thoreau (1962, *X*, 150) suggested that "if you are afflicted with melancholy [. . .], go to the swamp." He subscribed to the miasmatic theory of neither malaria, nor of melancholy unlike some of his contemporaries, nor to the theory of the four elements and humors in the dominant paradigm in mainstream western culture.

Thoreau upset the conventional view of the dominant cultural paradigm that swamps were poisonous by parodying it in his reference to the "rank and venomous luxuriance in this swamp" (Thoreau 1962, *IX*, 60). Even at the worst of times he could prescribe a swamp cure, such as when he was at Beck Stow's Swamp on July 18, 1852. Thoreau (1962, *IV*, 231) wrote in his *Journal* for that day and about this swamp that "when life looks sandy and barren, is reduced to its lowest terms, we have no appetite, and it has no flavor, then let me visit such a swamp as this, deep and impenetrable, where the earth quakes for a rod around you at every step, with its open water where swallows skim and twitter." When desire is diminished and life is dissatisfying, both figured here orally, the quaking zone of the swamp has depths and softness that the shallowness of its waters belie and that the depths of the lake cannot dream of. Thoreau values precisely those usual denigratory connotations in the dominant cultural paradigm that attach to the "depth," or horizontal extension and impenetrability of the swamp.

Thoreau also upsets the usual dissociation between swamp and city by seeing the city as swamp and so subverting the unfavorable connotations of the swamp as a place of darkness, disease and even death. Rather than the swamp, Thoreau saw "society," "civilization" and the modern city as bearers of disease, or perhaps more precisely he saw the modern city as swamp in the conventional sense of an uncanny and unhomely place of disease and horror, and saw simultaneously the swamp as canny and homely, as postmodern dwelling in the unconventional sense of a homely, but also wild, (homely *because* wild) place. Thoreau (1962, *II*, 47) wrote in his *Journal* in 1850 that he could "see less difference between a city and some dismallest swamp then formerly. It is a swamp too dismal and dreary, however, for me." The city for Thoreau is dismal, not the city. He upsets and overturns the conventional association between swamp and dismalness, and the city not being dismal. Although he would prefer the swamp as swamp over the city as swamp, he nevertheless goes on to make a finer distinction: "I would prefer even a more cultivated place, free from miasma and crocodiles." And wouldn't we all? He still prefers the swamp to the city, and the native quaking zone of the swamp where crocodiles live to the feral quaking zone of the city, such as its swampy slums (Giblett 2009, 1–54).

Perhaps Thoreau was wise to prefer "a more cultivated place" free of crocodiles, though this did not mean that he preferred the city to the swamp, especially as the former could be as dangerous in common parlance as the latter. The city as swamp for Thoreau (1982, 350) had its own diseases and horrors:

> Let us settle ourselves, and work and wedge our feet downward through the mud and slush of opinion, and prejudice, and tradition, and delusion, and appearance, that alluvion which covers the globe, through Paris and London, through New York and Boston and Concord, through church and state, through poetry and philosophy and religion, till we come to a hard bottom and rocks in place which we call *reality*.

The four cities and the town mentioned were founded in, or next to, and expanded into, swamps (for Paris, London, New York, and Boston, see Giblett 2016, chapters 3, 4, 9, and 10).

QUAKING ZONE

For Thoreau "there is a hard bottom everywhere," even "with the bogs and quick-sands of society" (Thoreau 1982, 568–589). The swamp, by contrast, for Thoreau (1962, *IX*, 394) writing in his *Journal* for May 31, 1857, is shallow and soft, the first birth of nature:

> That central meadow and pool in Gowing's Swamp is its very navel, *omphalos*, where the umbilical cord was cut that bound it to creation's womb. Methinks every swamp tends to have or suggests such an interior tender spot. The sphagnous crust that surrounds the pool is pliant and quaking, like the skin or muscles of the abdomen; you seem to be slumping into the very bowels of the swamp.[8]

In other words, slumping into the abject body of the earth. The swamp as creation's womb is the central tenet of Thoreau's of evolutionary biology and his conservation counter-creationist theology of swampy evolution. Gowing's Swamp, like the wetland of Wyman Meadow, dries up and gets wet during the seasonal cycle of the rising and falling waters of the underlying aquifer, which is typical of wetlands.

Another way in which Thoreau differs from the standard rhetoric of swamp-speak in which the swamp is a place of death and disease is that for him the swamp is the stuff of life and death. Indeed, for Thoreau in a letter to Emerson of March 11, 1842, "death is only the phenomenon of the individual or class. Nature does not recognise it" (Thoreau 1958, 64; 2013, 104). Moreover, for him in the same letter death is "a law and not an accident—It is as common as life [. . . T]he law of their [flowers'] death is the law of new life" (1958, 64; 2013, 105). The swamp, as with nature generally, upsets the hard and fast distinction between life and death. Thoreau (1993, 100–101; 1962, *XIV*, 109) inverts the morbid Christian orthodoxy of the line from the Anglican/Church of England *Book of Common Prayer* that "in the midst of life, we are in death" by maintaining how "in the midst of death, we are in life." In the middle of death in the swamp, we are in life. The swamp as marrow is constantly being renewed by the life-blood of the earth and constantly renews the bones of the body of the earth that give it structure. This is Thoreau's evolutionary biology and his conservation counter-theology of swamps in a nutshell.

The wetland of Wyman Meadow is next to his house site at Walden Pond and the wetland of Gowing's Swamp is within a mile (about a kilometer) of Thoreau's birthplace (see map of Concord, Mass, and Index, Thoreau 1962, *XIV*). Of the hundreds of places and sites he wrote about, Gowing's Swamp is the closest to his birthplace. Thoreau was a frequent visitor to Gowing's Swamp. He wrote about it repeatedly and drew two maps of it in his *Journal* where he described "its quaking ground" and "*navel* pool in the centre" (1962, *VI*, 467; *X*, 198 and *XIII*, 125). The wetland of Wyman's Meadow is close to his house site, just as the swamp itself is closest to his heart. Gowing's Swamp and Wyman Meadow were the womb and birthplace of American conservation, whereas "Walden Woods," including its Pond, were the "Cradle of American Conservation."[9] Swamps, marshes, and meadows were there at the beginning in the conception, and then later in the maturation, of Thoreau's conservation thinking.

The swamp is a native quaking zone of horror as Thoreau (1962, *VI*, 467) also suggested in his *Journal* for August 21, 1854, when he examined the middle of Gowing's Swamp and found an open pool of water nearly full of sphagnum and other plants, but he could not see them "on account of the danger of standing on the quaking ground [. . .] [and] of quaking sphagnum in which I sink eighteen inches in water [. . .] where I fear to break through." A quaking zone is a dangerous place that elicits fear and trembling because it not only mixes land and water creating mud and slime, but also because it threatens to engulf the whole body fully in its slimy depths, immerse the subject in the abject, and mark the surface of the body with mud.

Thoreau (1962, *VI*, 479) found these aspects again five days later while "picking our way over quaking meadows and swamps and occasionally slipping into the muddy batter midleg deep." The native quaking zone does not give a secure and stable place for standing upright, having a fixed point of view and observing the scene before one—preferably a pleasing prospect stretching before one as a subject with an object. The native quaking zone threatens to engulf the body in the slimy depths of the abject and return the subject to the watery womb from which it was born. Outside this zone, however, the view from the edge is another matter. Thoreau (1962, *XIV*, 301–302) writes on December 30, 1860, in the final volume of his *Journal* of "the open swamp [. . .] where the surface quakes for a rod around:" "there is no wilder and richer sight than is afforded from such a point of view of the edge of a [. . .] swamp." The native quaking zone affords visual pleasure from a distance, but provokes tactile horror up close and personal.

The native quaking zone is a bodily and earthly zone, the womb of the great goddess earth, or to use Mikhail Bakhtin's (Giblett 2008a, chapter 4) terms, the grotesque lower earthly and bodily stratum (see Giblett 1996, chapter 6). The native quaking zone of swamps and other wetlands can be a place of horror, whereas the feral quaking zone of polluted wastelands is a place of terror. Horror and terror, the two ruling passions of modernity (see Giblett 1996, chapter 2), preside over them. The quaking zone is a way of thinking about human beings, our bodies and minds, and the earth as connected through the possession of a common, contiguous and continuous area. No more so than when it comes to considering alligators and crocodiles. The native quaking zone of the tropical swamp is home to the alligator and crocodile. The alligator for Thoreau (1997, 283) "comes out of the mud with quakings of the earth."

The surface of the swamp is the soft spot of nature, even the breasts of the Great Mother of the swamps when Thoreau (1962, *IX*, 38–39) refers to "the soft open sphagnous centre of the swamp" as "these sphagnous breasts of the swamp—swamp pearls." The soft centre of the swamp is also related to the human body for Thoreau (1962, *X*, 262) as "the part of you that is wettest is fullest of life." Thoreau not only resists the dominant patriarchal cultural paradigm, but also supports the alternative matrifocal cultural paradigm. Unlike many of his contemporaries, such as Harriett Beecher Stowe, who figured slavery as a swamp, slavery for Thoreau (1962, *VI*, 365) was the part of the body politic fullest of death: "slavery [. . .] has no life. It is only a constant decaying and a death, offensive to all healthy nostrils." Slavery, in other words, is the dead black waters of a polluted waste wetland, unlike the recurring decaying, death, and rebirth of the living black waters of a healthy wild wetland fragrant to Thoreau's healthy sense of smell important in the alternative cultural paradigm. Thoreau the conservationist placed the natality of being born again into the ecosphere of the earthly household (*oikos*) in ecology and economy at the centre of his environmental philosophy. Thoreau the activist placed the natality of being born again into the public sphere at the centre of his political philosophy. Thoreau the political conservationist did both. Thoreau wetland conservationists placed the natality of being born again in swamps at the centre of his aquaterratological philosophy and practice. Wetland conservation is aquaterranatality.

One of the attractions of the swamp for Thoreau, especially in winter, was that here was a place on which no other "man" had left a trace, and so it was a place where Thoreau (1962, *VIII*, 99) could leave his mark on a tabula rasa: "I love to wade and flounder through the swamp now, these bitter cold days when the snow lies deep on the ground [. . .] to wade through the swamps, all snowed up, untracked by man." Unlike the snow of field, pond, or road, the snow of the swamp could remain untracked for a time in order to allow Thoreau (1962, *VIII*, 160, 167) to write his own message on its clean sheet, its "blank page," without fear of interruption or interference from fellow humans, especially citizens, those denizens of the city.

After wading around in a swamp, Thoreau (1962, *IX*, 42) felt like an explorer: "I seemed to have reached a new world, so wild a place [. . .] far away from human society. What's the need of visiting far-off mountains and bogs, if a half-hour's walk will carry me into such wildness and novelty." Thoreau explored swamps not just physically but also metaphysically. Indeed, he did not even need to go on a half-hour's walk visiting bogs to be carried into wildness: "It is in vain to dream of wildness distant from ourselves. There is none such. It is the bogs in our brain and bowels, the primitive vigor of Nature in us, that inspires that dream. I shall never find in the wilds of Labrador any greater wildness than in some recess in Concord, that is, than I import into it" (Thoreau 1962, *IX*, 43). Wild(er)ness is a cognitive, corporeal, and cultural experience, not a geographical category of (wet)land conservation or use, or lack of it, indigenous or industrial.

Thoreau saw the swamp explorer as a kind of Columbus of the new world of swamps not only without but also within. He asked rhetorically, "Is not our own interior white on the chart? black though it may prove, like the coast, when discovered." Black is not a region of despair for Thoreau, but for new discoveries. He then also exhorts his readers to "be a Columbus to whole new continents and worlds within you, opening new channels, not of trade, but of thought" (Thoreau 1982, 560). The interior is either a kind of swamp in winter, a frozen *tabula rasa*, to be explored, mapped, written upon and so colonized or a swamp in summer with its quaking surface that could be decolonized and demapped.

For Thoreau (1982, 376) it is the screech owls, or more precisely "their dismal scream," which best express his view of marshes and swamps as dialogic other:

> I love to hear their wailing [. . .] as if it were the dark and tearful side of music [. . .] They are the spirits, the low spirits and melancholy forebodings, of fallen souls that once in human shape night-walked the earth and did the deeds of darkness, now expiating their sins with their wailing hymns or threnodies in the scenery of their transgressions. They give me a new sense of the variety and capacity of that nature which is our common dwelling.

Nature has just as much capacity for "evil" as it, or "she," has for "good." Nature is not all goodness and light for Thoreau, but also has its dark and evil "side." Yet the owls are unlike the Stymphalian birds of the Hercules myth in that they are not a monstrous deviation from nature that define and maintain the norm by contrast, but are a part of nature (for Hercules and the Stymphalian birds, see Giblett 1996, 180–181).

Nature for Thoreau (1982, 376) is both "our common dwelling," our homely setting of steam rising from kettle and swamp, and "this vast, savage, howling mother of ours" from whose breast "we are so early weaned [. . .] to society." Nature for Thoreau, unlike for his contemporaries and the dominant cultural paradigm, is both homely and unhomely, canny and uncanny. It is both a place of goodness and light perhaps exemplified by the clear "eyes" of the lake and pond, and a place of life and death, light and dark represented by the "marrow" of the swamp. Thoreau's double vision, arguably postmodern *avant la lettre* ("before the name"), embraces and entertains both at once without any sense of contradiction between them. The swamp is not a place of melancholy and madness for Thoreau, but a place where melancholy and madness are mediated and alleviated.

The screech owls function for Thoreau as a kind of post-Christian "scapegoat" (or more precisely scape-screech owls) that instead of being driven off into the premodern wilderness to bear the sins of "men" away from civilization and the city, are part and parcel of the postmodern wilderness (or in this case more precisely the scape-wetland of the wet landscape), in which "men" can find the sacred and solace, can find refuge and sanctuary from the rigors and stresses of modern city life:

> I rejoice that there are owls. Let them do the idiotic and maniacal hooting for men. It is a sound admirably suited to swamps and twilight woods which no day illustrates, suggesting a vast and undeveloped nature which men have not

> recognised. They represent the stark twilight and unsatisfied thoughts which all have. All day the sun has shone on the surface of some savage swamp [. . .] but now a dismal and fitting day dawns, and a different race of creatures awakes to express the meaning of Nature there. (Thoreau 1982, 377)

The owls suggest a premodern, matrifocal wetland that has not yet been subject to a patriarchal, developmental and industrial-technological imperative, yet which is now subjected to that imperative in the very act of naming it as "vast and undeveloped" with its meanings expressed by owls.

The postmodern wetland is worlds away from the melancholic marshes and the slough of despond found in John Bunyan's *Pilgrims Progress*: "there can be no very black melancholy to him who lives in the midst of Nature and has his senses still" (Thoreau 1982, 382). The place par excellence in which to live literally in the middle of Nature, even up to one's chin, is the swamp. Given the difficulties the swamp poses for travel, especially by modern western means of transportation, it is the perfect place to still the senses and the limbs, and to allow the swamp to write on them, not as a tabula rasa, but as a responsive surface.

The slimy edge of the swamp for Thoreau is not the place from which to flee for the bright and sublimed city lights, but the place to live for the bright swamp lights of ignited marsh gases that do not lead to madness, but could even lead to Thoreau's ultimate goal: "unto a life which I call natural I would gladly follow even a will-o'-the-wisp through bogs and sloughs unimaginable, but no man nor firefly has shown me the causeway to it" (Thoreau 1982, 625).

Thoreau seems to be developing a conservation language within the alternative cultural paradigm that would counter the standard Romantic perception exemplified by Wordsworth that "unless Nature sympathizes with and speaks to us, as it were, the most fertile and blooming regions are barren and dreary," in other words, are a modern wasteland (Thoreau 1962, *X*, 252). The postmodern wetland, by contrast, is where Nature does not necessarily sympathize with us, nor we with it, but speaks to us, as the screech owls do, in the most fertile and blooming regions of the swamp. The swamp may be bare, but certainly not barren: "in swamps where there is only here and there an evergreen tree amid the quaking moss and cranberry beds, the bareness does not suggest poverty" (Thoreau 1982, 195). The bareness does not suggest barrenness but fertility. Swamp water is living.

The postmodern wetland may not be beautiful in the conventional sense of possessing appropriate qualities of form, texture, color, depth of field and point of view. Perhaps that is why it has been regarded in and by the dominant cultural paradigm as barren and dreary. If the wetland had been regarded as beautiful, perhaps its perceived uselessness would not have been held so badly against it. Perhaps if the wetland could now be regarded as beautiful, the fact that it is "useless" as its stands for agriculture and urban development would not matter so much. For Thoreau (1993, 144), "whatever we have perceived to be in the slightest degree beautiful is of infinitely more value to us than what we have only as yet discovered to be useful and to serve our purpose." The trouble with marshes and swamps is that they have been regarded in and by the dominant cultural paradigm as lacking both beauty *and* utility. This double lack has been held against them.

The swamp may lack the typical characteristics of beauty, but it does possess gradation that Thoreau saw as one of the fundamental aesthetic and ecological hallmarks of nature. In his journal entry for January26, 1858, he wrote of Clintonia Swamp that "nature loves gradation" (Thoreau 1962, *X*, 260). A couple of weeks later in his journal entry for February 13, he wrote of Cafferty's Swamp that "the swamp was variously shaded, or painted even, like a rug, with the sober colors running gradually into each other" (Thoreau 1962, *X*, 282). Rather than subjecting marshes and swamps to an aesthetic and utilitarian, even capitalist, imperative, perhaps it would be preferable to see marshes and swamps as fulfilling vital, ecological functions necessary for life on earth to be sustained. Nature not only loves gradation in color, but also gradation between land and water, life and death, light and darkness—living black waters (see Giblett 1996). Long live living black waters!

NOTES

1. Email to the author May 10, 2019.
2. According to the Bureau of the Ramsar Convention on Wetlands of International Importance (Ramsar Convention Bureau n.d.).
3. As I viewed the wetland of Forrestdale Lake as I wrote many of my own books from my desk in front of a window facing the lake for twenty-five years.
4. Email to the author May 10, 2019.
5. Perhaps needless to say, in the house I built beside a wetland my window did not overlook my woodpile. I much preferred to look at the wetland with affection, rather than at my woodpile, and loved to have it before my window (Giblett 2013c, 140–142). I oriented our house so that it looked over the wetland to the south (not over the woodpile on the east side facing the hot morning sun in summer).
6. For further discussion of Thoreau, theology and his view of religion as being and living at home in one's body, time and place, see Giblett (2018b, 138–140).
7. For further discussion of Thoreau's religion, see Balthrop-Lewis (2021).
8. Carl Bode (1982, 686) coyly or squeamishly excludes the first sentence and last phrase when he quotes this passage from Thoreau's journal.
9. As proclaimed in the subtitle of a map (Walden Woods Project, 1991).

Chapter 9

Farewell

Nature Writing and Black Swan Lake

Tracing the life of the plants and animals of Forrestdale Lake through the six seasons of the local indigenous people, the first part of *Black Swan Lake* (Giblett 2013a) presents a wetlands calendar over a yearly cycle of the rising, falling, and drying waters of this internationally important wetland in south-western Australia. The second part of this book considers issues and explores themes from the first part, including a cultural history of the seasons and the black swan.

Black Swan Lake is a book of nature writing and environmental history and philosophy arising from living in a particular place with other beings. The book is a guide to living simply and symbiotically with the earth in troubled times and places by making and maintaining a strong attachment and vital connection to a local place and its flora and fauna. Local places and their living processes sustain human and other life on this living earth. I built my own house by Forrestdale Lake in 1986 and lived in it for twenty-eight years. For a couple of years, I kept a nature journal from which the first part of *Black Swan Lake* is drawn. The present chapter of *Wetlands and Western Culture* discusses *Black Swan Lake* in the context of previous nature writing, such as Henry David Thoreau and Aldo Leopold. It and *Wetlands and Western Cultures* as a whole concludes with a renewed plea for living bio- and psycho-symbiotic livelihoods in bioregional home habitats of the living earth (Giblett 2011, chapter 12).

Like Henry David Thoreau, one of the founders of the American conservation movement (as discussed in the previous chapter), I built my own house by a body of water and wrote a nature journal. Unlike Thoreau, however, who lived by Walden Pond for a couple of years and kept a nature writing journal for more than twenty years which was published posthumously in its entirety, I lived by Forrestdale Lake for more than twenty-five years and I wrote a nature journal for a couple of years which was published as the first part of *Black Swan Lake: Life of a Wetland* (Giblett 2013a). Like Thoreau's sojourn and journal writing by Walden Pond which gave rise to *Walden* (Thoreau 1997), the classic of nature writing and environmental philosophy and the old testament of conservation, my living and journal writing by Forrestdale Lake resulted in *Black Swan Like*, a book of both nature writing and of environmental, natural, and cultural history and philosophy.

Similarly, like Aldo Leopold, another founder of the American conservation movement, whose renovation of a chicken shack in a sand county in Wisconsin and nature writing journal gave rise to the new testament of conservation in *A Sand County Almanac* (Leopold 1949) whose first and second parts are devoted to nature writing and the third part to essays on issues arising from the first part, such as the land ethics and conservation aesthetic, *Black Swan Lake* is divided into two parts along similar lines with the first part devoted to nature writing and the second part to essays on issues arising from the first part, such as the natural and cultural history of the black swan, the color black, Perth as black swan (or swamp) city and a cross-cultural history of the seasons.

Lawrence Buell, the author of *The Environmental Imagination: Thoreau, Nature Writing, and the Formation of American Culture* (Buell 1995), wrote an endorsement for *Black Swan Lake* in which he said that:

> *Black Swan Lake* is a unique synthesis of bioregional knowledge, historical and philosophical reflection, and autobiographical narrative with a distinctly personal stamp. I have learned a great deal from it and I am sure that other readers concerned with environmental ethics and *the arts of environmental imagination,* both within the academy and beyond, will feel the same. (my emphasis)

Buell concurs that nature writing is more than about a sense of place and more about the processes of place and their workings within the wider context of the bioregion. Nature writing is also arguably bioregional writing. It is also a creative environmental art neglected in the usual roll call of the creative arts and a form of "practice-led research" as *Black Swan Lake* indicates with chapters in the second part devoted to a cultural history of the black swan, the color black, the seasons, the history of Perth, and so on.

By developing and practicing a bioregional perspective in *Black Swan Lake,* I was only following in the footsteps of Henry David Thoreau who is perhaps the writer to have most precisely and passionately expressed the sense of the bioregion as home-habitat (as we saw in the previous chapter). In his *Journal* he wrote how:

> Here I am at home. In the bare and bleached crust of the earth I recognize my friend [. . .] the constant endeavour should be to get nearer and nearer *here* [. . .] A man dwells in his native valley like a corolla in its calyx, like an acorn in its cup. *Here*, of course, is all that you love, all that you expect, all that you are. Here is your bride elect, as close to you as she can be got. Here is all the best and all the worst you can imagine. (Thoreau 1962, *XI*, 275; his emphases)

Home is here. It is not there. Life is here; it is not elsewhere. Home here is the bioregion. It is not the narrow confines of one's quarter acre, or fifth of an acre, suburban block, or one's townhouse and its pocket–handkerchief garden, or one's apartment and its balcony. One's bioregion is home; home is the bioregion. For Thoreau (1962 *VI*, 121) also "the earth is all alive." For us, too, today, humans need to (re-)learn to live with the earth as living being.

Developing a better relationship between humans and the earth, people, and place, culture and nature is vital for trying to achieve environmental sustainability in the age of climate change and national and international disasters. Yet the concept of "the environment" not only implies separation between humans and the earth, but also a subject-object relationship of the drive mastery over, and enslaving of, the earth. A master-slave relationship between humans and the earth is hardly sustainable, especially if sustainability is defined simply as "enough for all forever," and if "all" includes all creatures of the earth, not just humans. The slave, by definition, does not have enough. Enslavement takes those who had enough and makes them the property of those who had enough, but end up with more by virtue (or more precisely, vice) of having slaves—who ends up with less than enough. Rather than environmental sustainability, a more intimate and reciprocal relationship of mutuality with the earth means providing enough for all, including humans and other creatures on the earth, and the earth, forever. Social justice entails mutuality with the earth, and vice versa, where the social is the community of all beings.

One of the most powerful ways in which the drive for mastery over the earth is exercised is via the discourses of nature. I define discourses as institutionalized ways of seeing, saying and doing. Discourses of nature include: "the environment" that separates a subject from its environs, and produces a master-slave relationship between them; natural history that objectifies nature in taxonomic grids, and sets up a subject-object relationship between natural historian and nature; resource extraction that sadistically takes good things from nature in the form of commodities and returns bad things to it in pollutants and wastes; scientific ecology that colonizes nature and relegates the machinations of the earth household to a secluded feminine sphere; political economy that treats nature as a common source of free goods; nature aesthetics (beauty, picturesqueness, and sublimity) that aestheticizes nature in the surface of landscape and that valorizes the sense of sight and denigrates the others; monumentalism and sanctuarism that preserve bits of nature in national parks or wilderness zones, and exploit the rest; and resource extraction that commodifies nature wherever it can everywhere.

The main priority of sanctuarism is the preservation of species and ecosystems. While this is certainly important in the face of declining biodiversity and degraded habitats in the age of climate change, a richer appreciation of the world we live in is necessary to make the earth a more habitable place for all living beings. To this end I propose sacrality as the counter and complement of mutuality to the mastery of sanctuarism, and call for a shift in emphasis from the latter to the former. Sacrality embraces a broader sense of local and global space and place imbued with *signifiance* whose boundary is the ecosphere rather than just the biosphere (*signifiance* is Julia Kristeva's term for embodied, nonsensical, and playful processes of cultural production). The ecosphere includes bioregions and home as well as cities and communications in the electromagnetosphere (or "spectrum") and orbital extra-terrestrial space of the living earth.

One of the central ways in which humans relate to the earth is through work. Work is the means by which human life is sustained by the earth. The concepts and categories of "the environment," natural history, scientific ecology, landscape aesthetics, and their associated practices in conservation landscapes and industrial land use work-over (if not overwork) nature (defined simply as a collective noun for land, living beings, air, water, energy, and planetary motion). By contrast, Australian Aboriginal Country, conservation counter-aesthetics, and symbiotic livelihood in a bioregion work (with) the earth as living being. Instead of nature aesthetics and conservation landscapes that privilege some sites over others, and the sense of sight over the others, in *People and Places of Nature and Culture* (Giblett 2011) I advocate a conservation counter-aesthetic that appreciates and values all places and senses. In place of modern, scientific ecology and political economy separated from each other on either side of the nature/culture divide, I propose a participatory, postmodern political ecology that promotes sustainability in the earth household of the ecosphere. Instead of ways of seeing, saying, and doing in relation to nature, I call for a way of being exemplified in Aboriginal Country (Giblett 2011, chapter 11).

"Nature" is a problematic term and there is a number of competing definitions and discourses of it. Nature has been split between the first nature of indigenous cultures and the second nature of "agri-urban" cultures. While the latter constructs a subject-object distinction and relationship between people and the earth, the former is predicated on an inter-subjective, even mutually abjective relationship. The subject-object relationship is evident in natural history, modern scientific ecology, nature aesthetics, and landscape gardening as well as in nature conservationism in national parks and wilderness areas and nature exploitation in industrial land use.

Abjects are to be found in Aboriginal Country and in bioregional and symbiotic livelihood ("abject" is Julia Kristeva's (1982) term for the mediating category between subject and object that makes both possible). The subject-object relationship is evident in natural history, modern scientific ecology, nature aesthetics, and landscape gardening, as well as in nature conservation in national parks and wilderness areas, and nature exploitation in industrial landuse. Abjects are to be found in Aboriginal Country and in bioregional and symbiotic livelihoods.

This split between first and second nature is gendered; nature has a gendered construction. Not only has nature been feminized, but it has been feminized in two contradictory ways: as the life-giving and death-dealing Great Mother, or Great Goddess, associated with the swamps, and as the benign and malign Mother Earth or Mother Nature affiliated with the fields. Culture likewise has been split between matrifocal and gylanic cultures in which the sexes were equal, and patriarchal and hierarchical ones in which men are dominant (culture is defined simply as a collective noun for arts and crafts [including horticultural, hunting, gathering, building dwellings, culinary, decorative, and other domestic arts and crafts], language and writing [as both trace and inscription], ritual and exchange). Both of these splits have been mapped over each other. Split culture equates with split nature: matrifocal and gylanic cultures are associated with the Great Mother, or Great Goddess, whereas hierarchical and patriarchal cultures are aligned with the Mother Earth or Mother Nature. Split culture and nature cut across and deconstruct a simplistic distinction and unresolved binary opposition between culture and nature.

The benign and malign Mother Earth or Mother Nature is also affiliated and aligned with the European landscape aesthetic in which the surface of the land is an object of visual consumption and the depths of the land are either an object of exploitation in production or an abject of repression as has occurred with wetlands as I argue in *Postmodern Wetlands* (Giblett 1996). The European landscape aesthetic, in turn, produced the ways in which Europeans and their settler diaspora saw and shaped the land through the perceptions and practices of the gentleman's park (and suburban enclave) estate, national parks and wilderness, mining and pastoralism, and the "Bush" of Australian mateship (as we saw in chapter 6; see also Giblett 2011, chapters 4–10).

The drive for mastery over the earth culminates in industrial land use that results in ruination of the earth. In what Bruno Latour calls "the New Climate Regime," "there is no longer any question of 'mastering' nature" (Latour 2017, 111–115). Rather, "nature" is mastering, or monstering, us. The new, anthropogenic climate regime rules—and it is *not* okay. This regime is what has come to be called "the Anthropocene" that came out of the economic politics of mercantile capitalism, industrial capitalism, enclosure of the commons, private property, the commodity market, and the feral quaking zone. It came, in short, out of the drive for, and the failed paradigm of, mastery (see Giblett 2011, chapter 1). The Anthropocene can be defined as the geological age of the laying and layering up and down of anthropogenic strata in the pollution of land, waters, and air, and the heating up of all three in the new, disrupted (and disruptive) arrangement of the four elements and the four seasons (see Giblett 2013c, chapter 18). The Anthropocene is also Anthropobscene, a laying bare of the wastes and pollutants that modern industrial capitalism and its technologies wanted the -ospheres (atmosphere, hydrosphere, biosphere, etc.) to hide. The repressed returns; the obscene comes out into the open, despite and because of the best efforts to keep it hidden.[1]

What Glenn Albrecht (2019) calls "the symbiocene" is the hoped-for age superseding the Anthropocene. The symbiocene is also the paradigm of the desire for mutuality with the Earth posed against the failed paradigm of the drive to mastery and the will to power of the anthropocentric over the Earth (see Giblett 2011, figure 2, 32–34 for these two paradigms in table form). The symbiocene can be defined as the geological age of the laying and layering up and down of bodies of land, waters, air, beings and things in biogenic and non-biogenic strata, and the inter-corporeal relationship of loving union between them in the feral and native quaking zones (see Giblett 2009, chapter 1). The symbiocene comes out of the economic politics of the commons, compassion for all beings, mutual aid, the carnivalesque marketplace (not the capitalist market) and the native quaking zone. The symbiocene comes out of its love for, and paradigm of, mutuality and psycho-symbiosis (see Giblett 2011, chapters 1 and 12).

In order to nurture and promote psycho-symbiosis and eco-mental health, I have undertaken previously an ecological psychoanalysis of the investments of desire and capital, yields of pleasure and profit, and relations of power and work in the mining and pastoral industries (Giblett 2011, chapters 9 and 10; 2019d, chapters 5 and 6). I argue for a move away from an emphasis on resource-exploitation, or greed and gluttony, to a relationship of generosity for gratitude, of respect for, reciprocity with and restoration of the earth. In more psychoanalytic terms, these chapters critique the oral and anal sadism of industrial land use in the mining and pastoral industries, and promote eco-mental health. The psychoanalytic ecology, participatory ecology and postmodern ecology developed throughout *People and Places of Nature and Culture* (Giblett 2011) address the personal, political, corporeal, cultural, and historical dimensions—the psychodynamics, economics, semiotics, and symbiotics – of our relationship with the earth. Working together they provide a platform for earthly mutuality and social justice. The final chapter celebrates a desire for dialogue and mutuality with the earth in bio- and psycho-symbiosis (Giblett 2011, chapter 12).

A participatory, postmodern, political ecology has a number of other features such as: deconstructing and decolonising the hierarchical privileging of the *polis* over the *oikos*, of the masculine public sphere over the feminized private sphere and biosphere; decolonising nature, including colonized regions of the human body; thinking critically about a communal sense of cultural and natural heritage; revaluing the spiritual interactions of human cultures with natural environments and earthly entities; and diagnosing the ecological symptoms in all discourses and theories (even when ecology is absent or ostensibly so). Its primary aim would be earthly mutuality in the interactions between the public and private spheres and the bio- and other -spheres that give and sustain life on this planet.

What constitutes nature and how our relationship to it is worked out is invariably couched in aesthetic terms. In any discussion of nature, aesthetics usually raises its ugly head! What we value in nature is usually what pleases us aesthetically. Conversely, what displeases us aesthetically is not usually valued culturally. Wetlands are a case in point as they are both aesthetically displeasing and culturally devalued. Wetlands are an impassable abject, the obverse of the impossible object of the sublime. The categories of aesthetics and landscape exercise mastery over nature. These categories are not ideologically neutral nor are they culturally universal but have colonized the lands and cultures of others.

Landscape is one of the central devices and means by which Europeans and their settler diasporas understand and relate to land. The European landscape aesthetic of the sublime, the picturesque and the beautiful denigrated wetlands (as we saw in chapters 3–5). The European landscape aesthetic of the sublime, the picturesque and the beautiful is a hierarchical taxonomy of landscapes in which, for instance, mountains are privileged over marshes, fields over fens. The European landscape aesthetic was part of the explorer's and settler's cultural baggage that they took with them and either found or recreated in the colonies with devastating consequences. Yet the indigenous inhabitants and owners of colonial land had lived sustainably with it for tens of millennia before.

The cultural history of nature can be traced beginning with the ancient Greek idea of nature as living organism and the modern European idea of nature as dead machine, through modern European landscape aesthetics of the sublime in vertiginous forms and the picturesque in the "pleasing prospects" of the gentleman's park estate, through their exploratory and colonising diasporas that saw and shaped the land in accordance with the European model, to the settler and industrial landscapes of national parks, wilderness, Australian bush, mining and pastoralism. Not only space (or more precisely three-dimensional, Euclidean space), but also time (or more precisely linear, chronological time) have been mastered through the enclosure and colonisation of pre-history, of the past, and foreclosure and colonisation of the future as implied by the double spatial and temporal meaning of "prospect." History has colonized pre- and pro-history.

Postmodern Wetlands: Culture, History, Ecology (Giblett 1996), focused, as the title and subtitle suggest, on a specific landform and on issues surrounding it. *People and Places of Nature and Culture* (Giblett 2011), by contrast, focused broadly on a range of different landscapes and issues surrounding them. Although the scope of this book is broad and its range extensive, it has clear spatial, temporal and conceptual parameters. The geographical trajectory traced in *People and Places* is from the European centre to the colonial periphery, from home to the unhomely, from the pastoral landscape of the gentleman's park estate at home and abroad in the pastoralist industry in the colonies and latterly in the entry statements of the enclave estates (both impacting fatally on the indigenous wetlands and woodlands of so-called "postcolonial" societies) to the "primitive wilderness" of indigenous peoples (and their conjuncture and clash).

A countervailing conceptual and temporal circle can be traced from nature as living organism through nature as dead matter back to the living earth; from the discourses (scientific and aesthetic) and economics (agricultural and industrial capitalist) of the drive for mastery to the practices of mutuality; from oral and anal sadism to bio- and psycho-symbiosis; from the cultural construction and discourse of nature as a way of seeing, saying and doing to Aboriginal Country as a way of being; from the rural, colonial, national and industrial to the indigenous, bioregional and ecospherical; from European city to Aboriginal Country; from history to pre-history; from timeline to time-circle; from Mother Nature to the Great Goddess; from sanctuarism to sacrality; from cultural nature to living earth; from landscape to land symbiotics; from mastery to mutuality. Mutuality with the earth entails living bio- and psycho-symbiotic livelihoods in bioregional home habitats of the living earth, including the living black waters of wetlands. Please help save them. Black waters live!

NOTE

1. What Freud and Schelling called, and how they defined, the uncanny; see Giblett (2019b, chapters 1 and 4).

Bibliography

AAP 2019. "Budj Bim Indigenous Eel Trap Site Added to World Heritage List." *The Guardian,* July 7. Accessed July 7, 2020. https://www.theguardian.com/australia-news/2019/jul/07/budj-bim-indigenous-eel-trap-site-added-to-world-heritage-list.

Abraham, Nicola and Maria Torok. 1984. "A Poetics of Psychoanalysis: 'The Lost Object - Me.'" *SubStance*, no. 43: 3–18.

Albrecht, Glenn. 2019. *Earth Emotions: New Words for a New World.* Ithaca, NY: Cornell University Press.

Annear, Robyn. 2014. *Bearbrass: Imagining Early Melbourne.* Melbourne: Black Inc.

Arendt, Hannah. 1996. *Love and Saint Augustine.* Edited by Joanna Vecchiarelli Scott and Judith Chelius Stark. Chicago: University of Chicago Press. First published in German in 1929.

Arnold, John. 1983. *The Imagined City: Melbourne in the Mind of its Writers.* North Sydney: George Allen and Unwin.

Arrowsmith, John. 1833. *Discoveries in Western Australia: From Documents Furnished to the Colonial Office by J.S Roe, Esqre. Surv. Genl.* [Map]. Accessed June 8, 2020. https://nla.gov.au/nla.obj-232547831/view.

Arrowsmith, John. 1848-49. *The Colony of Western Australia: From the Surveys of John Septimus Roe Esqr. Surveyor Genl. and from Other Official Documents in the Colonial Office and Admiralt.* [Map]. Accessed June 8, 2020. https://nla.gov.au/nla.obj-232548449/view.

Arsić, Branka. 2016. *Bird Relics: Grief and Vitalism in Thoreau.* Cambridge, MA: Harvard University Press.

Ashton, Paul. 2008. "Centennial Park." In *The Dictionary of Sydney.* Accessed February 29, 2020. https://dictionaryofsydney.org/entry/centennial_park.

Ashton, Paul, Kate Blackmore and Armanda Scorrano. 2013. *People's Park – Centennial Park: A History.* Ultimo, NSW: Halstead Press.

Attenbrow, Val. 2010. *Sydney's Aboriginal Past: Investigating the Archaeological and Historical Records,* second edition. Sydney, NSW: UNSW Press.

Australian Associated Press. 2020. "Western Australia's King Leopold Ranges renamed Wunaamin Miliwundi Ranges." *The Guardian,* July 4. Accessed July 4, 2020. https://www.theguardian.com/australia-news/2020/jul/03/western-australias-king-leopold-ranges-renamed-to-honour-aboriginal-heritage.

Balthrop-Lewis, Alda. 2021. *Thoreau's Religion: Walden Woods, Social Justice, and the Politics of Asceticism.* Cambridge: Cambridge University Press.

Barrell, John. 1972. *The Idea of Landscape and the Sense of Place 1730-1840: An Approach to the Poetry of John Clare.* Cambridge: Cambridge University Press.

Benjamin, Walter. 1977. *The Origin of German Tragic Drama.* Translated by John Osborne. London: New Left Books.

Benjamin, Walter 1999. *The Arcades Project.* Translated by Howard Eiland. Cambridge, MA: Belknap Press of Harvard University Press.

Benjamin, Walter. 2019. *Origin of the German Trauerspiel.* Translated by Howard Eiland. Cambridge, MA: Harvard University Press.

Benjamin, Walter and Gershom Scholem. 1992. *The Correspondence of Walter Benjamin and Gershom Scholem, 1932-1940.* Edited by Gershom Scholem. Translated by Gary Smith and Andre Lefevere. Cambridge, MA: Harvard University Press.

Benveniste, Émile. 2016. *Dictionary of Indo-European Concepts and Society.* Translated by Elizabeth Palmer. Chicago: Hau Books. First published in French in 1969.

Beowulf and the Finnesburg Fragment. 1950. Translated by J. Clark Hall. New edition completely revised with an introduction by C. L. Wrenn and with prefatory remarks by J. R. R. Tolkien. London: George Allen and Unwin.

Beowulf. 1999. Translated by Seamus Heaney. London: Faber and Faber.

Bertram, Nigel and Catherine Murphy. 2019. "The Swampy Lowlands of Melbourne." In *In Time with Water: Design Studies of 3 Australian Cities*, edited by Nigel Bertram and Catherine Murphy, 79–153. Crawley: University of Western Australia Publishing.

Bird, Rod. 2014. *A Short History of Wetlands in SW Victoria: The Landscape at Settlement, Impact of Agriculture, Major Wetlands near Hamilton, and Wetland Restoration.* Hamilton, Victoria: R. Bird.

Bode, Carl. 1982. "Epilogue by the Editor." In *The Portable Thoreau,* edited by Carl Bode, 683–696. New York: Viking Penguin.

Borch-Jacobson, Mikkel. 1989. *The Freudian Subject.* Translated by Catherine Porter. London: Macmillan.

Brady, Danielle and Jeffrey Murray. 2020. "The Cultural Significance of Wetlands: Perth's Lost Swamps to the Beeliar Wetlands." In *Australian Wetland Cultures: Swamps and the Environmental Crisis,* edited by John C. Ryan and Li Chen, 175–194. Lanham, MD: Lexington Books.

Bregman, Rutger. 2020. *Humankind: A Hopeful History.* Translated by Elizabeth Manton and Erica Moore. London: Bloomsbury.

Bride, Thomas Francis, editor. 1898. *Letters from Victorian Pioneers: Being a Series of Papers on the Early Occupation of the Colony, the Aborigines, etc. Addressed by Victorian Pioneers to his Excellency Charles Joseph La Trobe, Esq., Lieutenant*

Governor of the Colony of Victoria. Melbourne: Trustees of the Public Library of Victoria.

Brittain, Vera. 1978. *Testament of Youth: An Autobiographical Study of the Years 1900-1925*. London: Virago.

Brooks, Paul. 2014. *Speaking for Nature: The Literary Naturalists from Transcendentalism to the Birth of the American Environmental Movement*. Mineola, NY: Dover.

Brown, Andrew. 1993. "Orchids." In *Mountains of Mystery: A Natural History of the Stirling Range*, edited by Carolyn Thomson, Graham Hall and Gordon Friend, 61–70. Como, Western Australia: Department of Conservation and Land Management.

Buell, Lawrence. 1995. *The Environmental Imagination: Thoreau, Nature Writing, and the Formation of American Culture*. Cambridge, MA: Harvard University Press.

Bunyan, John 2008. *The Pilgrim's Progress*. Edited by Roger Pooley. London: Penguin. First published in 1678.

Burton, Robert. 1932. *The Anatomy of Melancholy*. Edited by Holbrook Jackson. London: Dent. First published in 1621.

Cannon, Michael. 1976. *The Land Boomers*. New Illustrated Edition. Melbourne: Thomas Nelson.

Carter, Paul. 1987. *The Road to Botany Bay: An Essay in Spatial History*. London: Faber.

Carter, Paul. 2005. *Mythform: The Making of* Nearamnew *at Federation Square*. Carlton, Victoria: Miegunyah Press of Melbourne University Press.

Cathcart, Michael. 2010. *The Water Dreamers: The Remarkable Story of our Dry Continent*. Melbourne: Text.

Cioran, E. M. 1995. *Tears and Saints*. Translated by Ilinca Zarifopol-Johnston. Chicago: University of Chicago Press. First published in Romanian in 1937.

Cioran, E. M. 2010. *A Short History of Decay*. Translated by Richard Howard. London: Penguin. First published in French in 1949.

Cioran, E. M. 2013. *The New Gods*. Translated by Richard Howard. Chicago: University of Chicago Press. First published in French in 1969.

Cioran, E. M. 2020. *The Trouble with Being Born*. Translated by Richard Howard. London: Penguin. First published in French in 1973.

Coode, John. 1879. "Melbourne Harbour Trust: Works of Improvement." Accessed May 1, 2015. https://viewer.slv.vic.gov.au/?entity=IE15504137.

Cotman, John Sell. ca 1835. "Drainage Mills in the Fens, Croyland, Lincolnshire." Accessed January 7, 2020. https://collections.britishart.yale.edu/vufind/Record/1666774.

Cotman, John Sell (attrib.). "St. Benet's Abbey." Accessed January 7, 2020. https://collections.britishart.yale.edu/vufind/Record/1666197.

Cox, Henry. 1866. "Victoria-Australia, Port Phillip. Hobson Bay and Yarra River Leading to Melbourne." Accessed May 1, 2015. https://viewer.slv.vic.gov.au/?entity=IE7192876&mode=browse.

Creed, Barbara. 1986. "Horror and the Monstrous Feminine." *Screen* 27, no. 1: 44–70.

Cunningham, Sophie. 2011. *Melbourne*. Sydney: New South.

Darby, Andrew. 2020. *Flight Lines: Across the Globe on a Journey with the Astonishing Ultramarathon Birds.* Crows Nest, NSW: Allen & Unwin.

Davis, Hubert. 1962. *The Great Dismal Swamp: Its History, Folklore and Science*, revised edition. Self-published.

Deleuze, Gilles and Felix Guattari. 1977. *Anti-Oedipus: Capitalism and Schizophrenia.* Translated by Roger Hurley, Mark Seem and Helen Lane. New York: Viking Press.

Department of Environment, Land, Water and Planning (DELWP). 2019. *Wetlands of the Greater Grampians.* Accessed March 25, 2020. https://www.water.vic.gov.au/waterways-and-catchments/our-catchments/ococ/projects/wetlands-of-the-greater-grampians.

De Wint, Peter. 1815a. "A Cornfield." Accessed January 7, 2020. http://collections.vam.ac.uk/item/O15129/a-cornfield-oil-painting-peter-de-wint/.

De Wint, Peter. 1815b. "Lincolnshire Cornfield near Horncastle." Accessed January 7, 2020. https://artuk.org/discover/artworks/lincolnshire-cornfield-near-horncastle-81680/view_as/grid/search/keyword:lincolnshire-landscape--actor:de-wint-peter-17841849/page/1.

De Wint, Peter. n.d. "Mrs Peter de Wint." Accessed online, January 7, 2020. https://www.artuk.org/discover/artworks/mrs-peter-de-wint-17911866-81833.

Di Palma, Vittoria. 2014. *Wasteland: A History.* New Haven, CT: Yale University Press.

Douglas, Mary. 1966. *Purity and Danger: An Analysis of the Concepts of Danger and Taboo.* London: Routledge.

Dovey, Kim and Ron Jones. 2018. "Introduction." In *Urban Choreography: Central Melbourne 1985–*, edited by Kim Dovey, Rob Adams and Ron Jones, 9–12. Carlton, Victoria: Melbourne University Press.

Eagleton, Terry. 1986. *William Shakespeare.* Oxford: Basil Blackwell.

Eggleston, C. L. 2012. "Wilderness, Desert." In *Dictionary of the Old Testament: Prophets,* edited by Mark J. Boda and J. Gordon McConville, 843–847. Downers Grove, IL: InterVarsity Press.

Eilenberger, Wolfram. 2020. *Time of the Magicians: Wittgenstein, Benjamin, Cassirer, Heidegger and the Decade that Reinvented Philosophy.* Translated by Shaun Whiteside. New York: Penguin.

Elborough, Travis. 2016. *A Walk in the Park: The Life and Times of a People's Institution*, London: Jonathan Cape.

Eldershaw, Flora. 1952. "The Landscape Writers." *Meanjin* 3, no. 11: 215–228.

Eldershaw, M. Barnard. 1939. *My Australia.* London: Jarrolds.

Falconer, Delia. 2010. *Sydney.* Sydney: New South Books.

Fanon, Franz. 1967. *The Wretched of the Earth.* Translated by Constance Farrington. Harmondsworth: Penguin. First published in French in 1965.

Ferber, Ilit. 2013. *Philosophy and Melancholy: Benjamin's Early Reflections on Theater and Language.* Stanford: Stanford University Press.

Flannery, Tim, editor. 1999. *The Birth of Sydney.* Melbourne: Text.

Flannery, Tim. 1999. "The Sandstone City." In *The Birth of Sydney*, edited by Tim Flannery, 1–43. Melbourne: Text.

Flannery, Tim, editor. 2002. *The Birth of Melbourne.* Melbourne: Text.

Flannery, Tim. 2002. "The Passing of Birrarang." In *The Birth of Melbourne,* edited by Tim Flannery, 1–23. Melbourne: Text.

Fletcher, Brian. 1989. "Sydney: A Southern Emporium." In *The Origins of Australia's Capital Cities*, edited by Pamela Statham, 52–78. Cambridge: Cambridge University Press.

Forster, Edward M. 1936. *A Passage to India*. Harmondsworth: Penguin.

Foucault, Michel. 1977. *Discipline and Punish: The Birth of the Prison.* Translated by A. Sheridan. Harmondsworth: Penguin. First published in French in 1975.

Foucault, Michel. 1991. "Governmentality." In *The Foucault Effect: Studies in Governmentality with Two Lectures and an Interview* [...], edited by Graham Burchell, Colin Gordon and Peter Miller, 87–104. London: HarvesterWheatsheaf.

Freud, Sigmund. 1984. *On Metapsychology: The Theory of Psychoanalysis*, Pelican Freud Library 11. Harmondsworth: Penguin.

Freud, Sigmund. 2005. *On Murder, Mourning and Melancholia.* Translated by Shaun Whiteside. London: Penguin.

Fritzell, Peter. 1978. "American Wetlands as Cultural Symbol: Places of Wetlands in American Culture." In *Wetland Functions and Values: The State of Our Understanding*, edited by P. E. Greeson, J. R. Clark and J. E. Clark, 523–534. Minneapolis, MN: American Water Resources Association.

Esbjorn-Hargens, Sean and Michael Zimmerman. 2009. *Integral Ecology: Using Multiple Perspectives on the Natural World.* Boston: Shambhala.

Gabaldon, Diana. 2009. *An Echo in the Bone: A Novel.* London: Orion.

Gammage, Bill. 2011. *The Biggest Estate on Earth: How Aborigines made Australia.* Sydney: Allen & Unwin.

Gandy, Matthew. 2014. *The Fabric of Space: Water, Modernity, and the Urban Imagination.* Cambridge, MA: The MIT Press.

Garth, John 2003. *Tolkien and the Great War: The Threshold of Middle-Earth.* New York: Houghton Mifflin.

Garth, John. 2006. "Frodo and the Great War." In *Lord of the Rings, 1954-2004: Scholarship in Honor of Richard E. Blackwelder*, edited by Wayne G. Hammond and Cristina Scull, 41–56. Milwaukee, WI: Marquette University Press.

Geist, Johann. 1983. *Arcades: The History of a Building Type*. Cambridge, MA: The Massachusetts Institute of Technology.

Giblett, Rod. 1992. "Philosophy (and Sociology) in the Wetlands: The S(ub)lime and the Uncanny." *New Formations* no. 18: 142–159.

Giblett, Rod. 1993a. "Cities and Swamp Settling: Decolonizing Wetlands." *Span: Journal of the South Pacific Association of Commonwealth Literature and Language Studies* 36, no. 1: 285–301.

Giblett, Rod. 1993b. "Kings in Kimberley Watercourses: Sadism and Pastoralism." *Span: Journal of the South Pacific Association of Commonwealth Literature and Language Studies* 36, no. 2: 541–559.

Giblett, Rod. 1993c. "Outer City Sanctuary: Forrestdale Lake." *Landscope* 8, no. 3: 35–38.

Giblett, Rod. 1996. *Postmodern Wetlands: Culture, History, Ecology.* Edinburgh: Edinburgh University Press.

Giblett, Rod. 2006. *Forrestdale: People and Place.* Bassendean: Access Press.

Giblett, Rod. 2007. "Black and white water." In *Fresh water: New Perspectives on Water in Australia,* edited by Emily Potter, Alison Mackinnon, Stephen McKenzie and Jennifer McKay, 31–43. Carlton: Melbourne University Press.

Giblett, Rod. 2008a. *The Body of Nature and Culture.* Basingstoke: Palgrave Macmillan.

Giblett, Rod. 2008b. *Health Recovery: The* Taoist Tai Chi™ *Way.* London: Shepheard-Walwyn.

Giblett, Rod. 2008c. *Sublime Communication Technologies.* Basingstoke: Palgrave Macmillan.

Giblett, Rod. 2009. *Landscapes of Culture and Nature.* Basingstoke: Palgrave Macmillan.

Giblett, Rod. 2011. *People and Places of Nature and Culture.* Bristol: Intellect Books.

Giblett, Rod. 2012. "Nature Is Ordinary Too: Raymond Williams as the Founder of Ecocultural Studies." *Cultural Studies* 26, no. 6: 922–933.

Giblett, Rod. 2013a. *Black Swan Lake: Life of a Wetland.* Bristol: Intellect Books.

Giblett, Rod. 2013b. "A City 'Set in Malarial Lakeside Swamps': Toronto and Ashbridge's Bay Marsh." *Topia: Canadian Journal of Cultural Studies* no. 29: 113–132.

Giblett, Rod. 2013c. *Traces: Of an Active and Contemplative Life, 1983-2013.* Champaign, Illinois: Common Ground.

Giblett, Rod. 2014. *Canadian Wetlands: Places and People.* Bristol: Intellect Books.

Giblett, Rod. 2016a. *Cities and Wetlands: The Return of the Repressed in Nature and Culture.* London: Bloomsbury.

Giblett, Rod. 2016b. "Lost and Found Wetlands of Melbourne." *Victorian Historical Journal* 87, no. 1: 134–155.

Giblett, Rod. 2018a. *The Dragon and Saint George: A Fairy Tale Novella.* Cambridge: Pegasus/Elliott Mackenzie.

Giblett, Rod. 2018b. *Environmental Humanities and Theologies: Ecoculture, Literature and the Bible.* London: Routledge.

Giblett, Rod. 2018c. *Tales of Two Dragons.* London: Austin Macauley.

Giblett, Rod. 2019a. *Black Swan Saga.* Sharjah: Austin Macauley.

Giblett, Rod. 2019b. *Environmental Humanities and the Uncanny: Ecoculture, Literature and Religion.* London: Routledge.

Giblett, Rod. 2019c. *The Oxbridge Book of Dragons.* London: Olympia.

Giblett, Rod. 2019d. *Psychoanalytic Ecology: The Talking Cure for Environmental Illness and Health.* London: Routledge.

Giblett, Rod. 2020a. "Artist and Swamp: Wetlands in Australian Painting and Photography." In *Australian Wetland Cultures: Swamps and the Environmental Crisis,* edited by John Charles Ryan and Li Chen, 51–70. Lanham, MD: Lexington Books.

Giblett, Rod. 2020b. "Epilogue: Twenty-Five Years of Wetland Studies in the Humanities." In *Australian Wetland Cultures: Swamps and the Environmental Crisis,* edited by John Charles Ryan and Li Chen, 227–239. Lanham, MD: Lexington Books.

Giblett, Rod. 2020c. *Modern Melbourne: City and Site of Nature and Culture.* Bristol: Intellect Books.

Giblett, Rod. 2020d. *New Lives of the Saints: Twelve Environmental Apostles.* Lanham, MD: Hamilton Books.

Giblett, Rod. 2020e. "Rainbow Serpent Anthropology, or Rainbow Spirit Theology, or Swamp Serpent Sacrality and Marsh Monster Maternity?" In *Australian Wetland Cultures: Swamps and the Environmental Crisis,* edited by John Charles Ryan and Li Chen, 33–50. Lanham, MD: Lexington Books.

Giblett, Rod. 2021. *Black Swan Song: Life and Work of a Wetland Writer.* Lanham, MD: Hamilton Books.

Giblett, Rod. Forthcoming. *Swamp Deaths: Collected Cold Cases and Other Marshy Mysteries.*

Giblett, Rod and David James. 2009. "Anstey-Keane Botanical Jewel." *Landscope* 24, no. 4: 41–44.

Giblett, Rod and Hugh Webb. 1993. "Western Australian Wetlands: Living Water or Useless Swamps?" *Habitat* 21, no. 3: 30–36.

Giblett, Rod and Hugh Webb, editors. 1996. *Western Australian Wetlands: The Kimberley and South-West.* Perth: Black Swan Press/Wetlands Conservation Society (Inc.).

Giblett, Rod and Hugh Webb, 1996. "Living Water or Useless Swamps?" In *Western Australian Wetlands: The Kimberley and South-West*, edited by Rod Giblett and Hugh Webb, 1–9. Perth: Black Swan Press/Wetlands Conservation Society (Inc.).

Giblett, Rod and Juha Tolonen. 2012. *Photography and Landscape.* Bristol: Intellect Books.

Golden Gate National Parks Conservancy. n.d. "Crissy Field." Accessed November 26, 2017. https://www.parksconservancy.org/programs/crissy-field-center.

Grann, David. 2017. *Killers of the Flower Moon: Oil, Money, Murder and the Birth of the FBI.* London: Simon and Schuster.

Grimwade, Stephen, editor. 2009. *Literary Melbourne: A Celebration of Writing and Ideas.* Prahran: Hardie Grant.

Gunditjmara People, the, with Wettenhall, Gib. 2010. *The People of Budj Bim: Engineers of Aquaculture, Builders of Stone House Settlements and Warriors Defending Country.* Heywood, Victoria: em PRESS Publishing for the Gunditj Mirring Traditional Owners Aboriginal Corporation.

GWM Water. n.d. *Lake Wartook.* Accessed March 25, 2020. https://www.gwmwater.org.au/using-lakes-and-reservoirs/our-reservoirs/lake-wartook.

Hall, Graham, Sylvia Leighton and Ray Wills. 1993. "European Exploration and History." In *Mountains of Mystery: A Natural History of the Stirling Range*, edited by Carolyn Thomson, Graham Hall and Gordon Friend, 35–40. Como, Western Australia: Department of Conservation and Land Management.

Hampl, Patricia. 2006. "Meadow." In *Home Ground: Language for an American Landscape,* edited by Barry Lopez and Debra Gwartney, 223–224. San Antonio, TX: Trinity University Press.

Haraway, Donna. 1992. "The Promises of Monsters: A Regenerative Politics for Inappropriate/d Others." In *Cultural Studies*, edited by Larry Grossberg, Cary Nelson and Paula Treichler, 295–337. New York: Routledge.

Hargreaves Associates. n.d. "Crissy Field." Accessed November 26, 2017. http://www.hargreaves.com/work/crissy-field/.

Heritage Council Victoria. 2005. "The Grampians, Southern Grampians Shire." *Victorian Heritage Database*. Accessed March 27, 2020. https://vhd.heritagecouncil.vic.gov.au/places/70397.

Hills, E. Sherbon. 1975. *The Physiography of Victoria: An Introduction to Geomorphology*. Melbourne: Whitcomb & Tombs. First published in 1940.

Hillsdon, Mark. 2020. 'Dry Run: The Wet Farming Experiment that could Sow Seeds for Future Crops.' *The Guardian*, November 12. Accessed November 13, 2020. https://www.theguardian.com/environment/2020/nov/12/dry-run-the-wet-farming-experiment-that-could-sow-seeds-for-future-crops-aoe.

Hinkson, Melinda. 2001. *Aboriginal Sydney: A Guide to Important Places of the Past and Present*. Canberra: Aboriginal Studies Press.

Hopkins, Gerard Manley. 1953. *Poems and Prose*. Harmondsworth: Penguin.

Hume, Fergus. 1999. *The Mystery of a Hansom Cab*. Melbourne: Text. First published in 1886.

Huxley, Thomas. 1989. "Struggle for Existence." In *Mutual Aid: A Factor of Evolution*, edited by Peter Kropotkin, 329–341. Montreal: Black Rose.

Jones, Ron. 2018. "Melbourne, Sung as It Were a New Song.' In *Urban Choreography: Central Melbourne 1985–*, edited by Kim Dovey, Rob Adams and Ron Jones, 85–142. Carlton, Victoria: Melbourne University Press.

Karskens, Grace. 2009. *The Colony; A History of Early Sydney*. Sydney: Allen & Unwin.

Keaney, Magdalene. 1998. "Eugene von Guérard 1811-1901: 'Mount William from Mount Dryden, Victoria, 1857.'" In *New Worlds from Old: 19th Century Australian and American Landscapes*, edited by Elizabeth Johns, Andrew Sayers and Elizabeth Kornhauser, 155. Canberra: National Gallery of Australia/Hartford, CT: Wadsworth Atheneum.

Keighery, Greg and John Beard. 1993. "Plant Communities." In *Mountains of Mystery: A Natural History of the Stirling Range*, edited by Carolyn Thomson, Graham Hall and Gordon Friend, 43–59. Como, Western Australia: Department of Conservation and Land Management.

Kiddle, Margaret. 1961. *Men of Yesterday: A Social History of the Western District of Victoria 1834-1890*. Melbourne: Melbourne University Press.

Kingsley, Charles. 1873. "The Fens." In *Prose Idylls: New and Old*, 93–131. London: Macmillan.

Kingsnorth, Paul. 2014. *The Wake*. London: Unbound.

Koori Heritage Trust. 1991. *Massacre Map*. Accessed April 8, 2020. https://cv.vic.gov.au/media/oldmedia/5755/massacre_File0001.jpg.

Korporaal, Glenda. 1991. "Merchants and Milestones." In *Sydney: Australia's First City*, edited by Edna Carew, 28–37. Double Bay, NSW: Focus Press.

Kristeva, Julia. 1982. *Powers of Horror: An Essay on Abjection.* Translated by Leon Roudiez. New York: Columbia University Press. First published in French in 1980.

Kristeva, Julia. 1989. *Black Sun: Depression and Melancholia.* Translated by Leon Roudiez. New York: Columbia University Press. First published in French in 1989.

Kristeva, Julia. 2001. *Hannah Arendt.* Translated by Ross Guberman. New York: Columbia University Press. First published in French in 1999.

Landscape Architecture Foundation. n.d. "Shanghai Houtan Park." Accessed November 26, 2017. https://www.landscapeperformance.org/case-study-briefs/shanghai-houtan-park.

Latour, Bruno. 2017. *Facing Gaia: Eight Lectures on the New Climate Regime.* Cambridge: Polity.

Lawrence, David Herbert. 1950. *Kangaroo.* Harmondsworth: Penguin. First published in 1923.

Leighton, Sylvia. 1993. "Aboriginal Culture." In *Mountains of Mystery: A Natural History of the Stirling Range*, edited by Carolyn Thomson, Graham Hall and Gordon Friend, 27–34. Como, Western Australia: Department of Conservation and Land Management.

Leopold, Aldo. 1949. *A Sand County Almanac and Sketches Here and There.* New York: Oxford University Press.

Lertzman, Renee. 2013. "The Myth of Apathy: Psychoanalytic Explorations of Environmental Subjectivity." In *Engaging with Climate Change: Psychoanalytic and Interdisciplinary Perspectives,* edited by Sally Weintrobe, 117–133. London: Routledge.

Lertzman, Renee. 2015. *Environmental Melancholia: Psychoanalytic Dimensions of Engagement.* London: Routledge.

Logan, Tyne and John Dobson. 2020. "Western Australia Bushfires Devastate the Stirling Ranges — One of the World's Richest Biodiversity Hotspots." *Australian Broadcasting Corporation News,* January 7. Accessed August 5, 2020. https://www.abc.net.au/news/2020-01-07/bushfires-cause-stirling-ranges-biodiversity-to-be-devastated/11844516.

McDowall, Carolyn. 2014. "Eugene Von Guérard – Colonial Artist, Nature Revealed." Accessed September 6, 2018. https://www.thecultureconcept.com/national-gallery-eugene-von-Guérard-nature-revealed.

Macfarlane, Robert. 2008. "Ghost Species." *Granta: The Magazine of New Writing,* no. 102*: The New Nature Writing,* 109–128.

Martin, Daniel Jan and Geoffrey London. 2019. "The Deep Waters of Perth." In *In Time with Water: Design Studies of 3 Australian Cities*, edited by Nigel Bertram and Catherine Murphy, 157–215. Crawley: University of Western Australia Publishing.

Martin, Ged. 1989. "The Founding of New South Wales." In *The Origins of Australia's Capital Cities*, edited by Pamela Statham, 37–51. Cambridge: Cambridge University Press.

Marx, Karl. 1976. *Capital: A Critique of Political Economy, Volume One.* Translated by Ben Fowkes. London: Penguin Books in association with *New Left Review.* First published in German in 1887.

Massola, Aldo. 1968. *Bunjil's Cave: Myths, Legends and Superstitions of the Aborigines of South-East Australia.* Melbourne: Lansdowne.
Massy, Charles. 2020. *Call of the Reed Warbler: A New Agriculture, A New Earth.* St Lucia, Queensland: Queensland University Press.
Mayes, Christopher. 2018. *Unsettling Food Politics: Agriculture, Dispossession and Sovereignty in Australia.* Lanham, MD: Rowman & Littlefield.
Mbembe, Achille. 2019. *Necropolitics.* Translated by Steven Corcoran. Durham, NC: Duke University Press.
Mbembe, Achille. 2021. *Out of the Dark Night: Essays on Decolonization.* New York: Columbia University Press.
McCrae, George Gordon. 1912. "Some Recollections of Melbourne in the 'Forties.'" *The Victorian Historical Magazine* no. 7: 114–136.
Menard, Andrew. 2018. *Learning from Thoreau.* Athens, GA: University of Georgia Press.
Milne, A. A. 1958. *The World of Pooh Containing Winnie-the-Pooh and the House at Pooh Corner.* London: Methuen.
Mitchell, Thomas. 1839. *Three Expeditions into the Interior of Eastern Australia: With Descriptions of the Recently Explored Region of Australia Felix...* Accessed April 8, 2020. http://gutenberg.net.au/ebooks/e00036.html#CHAPTER%203.14.
Mules, Warwick. 2014. *With Nature: Nature Philosophy as Poetics through Schelling, Heidegger, Benjamin and Nancy.* Bristol: Intellect Books.
National Gallery of Victoria. n.d. "Eugene von Guérard, 'Tower Hill,' 1855." Accessed September 6, 2018. www.ngv.vic.gov.au/essay/eugene-von-guerard-nature-revealed/3/.
National Library of Australia. n.d. "The Old Tank Stream, Sydney." Accessed April 6, 2020. https://catalogue.nla.gov.au/Record/339764?
National Museum of Australia. n.d. "The Old Tank Stream." Accessed April 6, 2020.
Nature Conservancy Australia. 2020. "Exploring Gayini - Nari Nari Country." Accessed April 6, 2020. https://www.natureaustralia.org.au/what-we-do/our-priorities/land-and-freshwater/land-freshwater-stories/gayini/.
Nolan, Sidney. 1955. "Carcase in Swamp." Tate Gallery, London. Accessed June 8, 2020. https://www.tate.org.uk/art/artworks/nolan-carcase-in-swamp-t03555.
North, Marianne. 1893. *Recollections of a Happy Life: Being the Autobiography of Marianne North, Volume II.* London: Macmillan.
North, Marianne. n.d. "Australasia." *Marianne North Online Gallery Kew Royal Botanic Gardens.* Accessed June 23, 2020. http://www.kew.org/mng/gallery/australasia.html.
NSW National Parks and Wildlife Service. n.d. "Macquarie Marshes Nature Reserve." Accessed October 27, 2019. https://www.nationalparks.nsw.gov.au/visit-a-park/parks/macquarie-marshes-nature-reserve.
Obeng-Odoom, Franklin. 2021. *The Commons in an Age of Uncertainty: Decolonizing Nature, Economy, and Society.* Toronto: University of Toronto Press.
Olver, Rob and Stuart Olver. 1998. *Dawn till Dusk: In the Stirling and Porongorup Ranges.* Nedlands: Tuart House of the University of Western Australia Press.

Ostrom, Elinor. 1990. *Governing the Commons: The Evolution of Institutions for Collective Action.* Cambridge: Cambridge University Press.

Otto, Kristen. 2005. *Yarra: A Diverting History.* Melbourne: Text.

Otto, Rudolf. 1950. *The Idea of the Holy: An Inquiry into the Non-Rational Factor in the Idea of The Divine and Its Relation to the Rational,* second edition. Translated by John W. Harvey. Oxford: Oxford University Press.

Owen, Chris. 2019. "The Pinjarra Massacre: It's Time to Speak the Truth of this Terrible Slaughter." *The Guardian,* November 18. Accessed July 4, 2020. https://www.theguardian.com/australia-news/2019/nov/18/the-pinjarra-massacre-its-time-to-speak-the-truth-of-this-terrible-slaughter.

Owen-Fisher, Becky and Nick Tearle. 2017. *Standing High out of Shrunken Peat.* NP: Compass Publishing.

Parks Victoria. 2003. *Grampians National Park Management Plan.* Accessed March 25, 2020. https://www.google.com/search?client=safari&rls=en&q=grampians+national+park+management+plan&ie=UTF-8&oe=UTF-8.

Parks Victoria. 2018. *Grampians National Park Visitor Guide.* Melbourne: Parks Victoria.

Pascoe, Bruce. 2014. *Dark Emu, Black Seeds: Agriculture or Accident.* Broome: Magabala Books.

Pastoreau, Michel. 2001. *Blue: The History of a Color.* Princeton: Princeton University Press.

Pensky, Max. 2001. *Melancholy Dialectics: Walter Benjamin and the Play of Mourning.* Amherst, MA: University of Massachusetts Press.

Pepys, Samuel. 1972. *The Diary, Volume IV, 1663.* Edited by Robert Latham and William Matthews. London: G. Bell.

Picon, Antione. 2000. "Anxious Landscapes: From the Ruin to the Rust." Accessed September 29, 2006. http://www.gsd.harvard.edu/people/faculty/picon/texts.html.

Piketty, Thomas. 2020. *Capital and Ideology.* Translated by Arthur Goldhammer. Cambridge, MA: Belknap Press of Harvard University Press.

Platt, Edward. 2019. *The Great Flood: Travels through a Sodden Landscape.* London: Picador.

Potter, Emily. 2019. *Writing Belonging at the Millennium: Notes from the Field on Settler-Colonial Place.* Bristol: Intellect Books.

Powell, J. M. 1970. *The Public Lands of Australia Felix: Settlement and Land Appraisal in Victoria 1834-91 with Special Reference to the Western Plains.* Melbourne: Oxford University Press.

Presland, Gary, editor. 1977. "Journals of GA Robinson, March-May 1841." *Records of the Victorian Archaeological Survey* no. 6. Melbourne: Ministry for Conservation Publication.

Pryor, Francis. 2019. *The Fens: Discovering England's Ancient Depths.* London: Head of Zeus.

Pullin, Ruth. 2011. *Eugene von Guérard: Nature Revealed.* Melbourne: National Gallery of Victoria.

Pullin, Ruth. 2018. *The Artist as Traveller: The Sketchbooks of Eugene von Guérard.* Ballarat: Art Gallery of Ballarat.

Pullman, Phillip. 2011. *His Dark Materials: Northern Lights, The Subtle Knife, The Amber Spyglass*. London: Everyman's Library.
Rajnai, Miklos, editor. 1982. *John Sell Cotman 1782-1842*. London: Herbert Press.
Ramsar Convention Bureau. n.d. "The Importance of Wetlands." Accessed May 1, 2015. https://www.ramsar.org/about/the-importance-of-wetlands.
Readfearn, Graham. 2020. "NSW Makes Record Land Purchase for New National Park." *The Guardian,* June 27. Accessed July 2, 2020. https://www.theguardian.com/environment/2020/jun/27/nsw-makes-record-land-purchase-for-new-national-park
Richardson, Jr, Robert D. 1993. "Introduction: Thoreau's Broken Task." In Henry David Thoreau, *Faith in a Seed*, edited by Bradley Dean, 3–17. Washington, DC: Island Press/Covelo, CA: Shearwater Books.
Robinson, George Augustus. 1998. *The Journals of* [...] *Chief Protector, Port Phillip Aboriginal Protectorate, Volume Two: 1 October 1840 – 31 August 1841.* Edited by Ian D. Clark. Melbourne: Heritage Matters.
Roe, John Septimus. 2014. *The Western Australian Explorations of* [...] *1829-1849.* Edited by Marion Hercock. Carlisle, Western Australia: Hesperian Press.
Royal Historical Society of Victoria. 2020. "The Swamp Vanishes." Accessed May 1, 2020. https://www.historyvictoria.org.au/event/exhibition-the-swamp-vanishes/.
Ruzicka, Annette. 2020. "'Kakadu of the South': 88,000 Hectares of NSW Wetland Handed Back to Nari Nari." *The Guardian,* April 6. Accessed April 6, 2020. https://www.theguardian.com/australia-news/gallery/2020/apr/06/kakadu-of-the-south-88000-hectares-of-nsw-wetland-handed-back-to-nari-nari.
Ryan, John Charles. 2012. *Green Sense: The Aesthetics of Plants, Place and Language.* Oxford: TrueHeart Press.
Ryan, John Charles. 2014. "Overland from Perth to Albany: A Journey of Botanical Intrigue." *Australian Garden History* 25, no. 3: 4–9.
Ryan, John C. and Li Chen. 2020. "Introduction to Australian Wetland Cultures: Thinking About (and with Swamps)." In *Australian Wetland Cultures: Swamps and the Environmental Crisis,* edited by John C. Ryan and Li Chen, 7–31. Lanham, MD: Lexington Books.
Ryan, John C., Danielle Brady and Chris Kueh. 2020. "Where Fanny Balbuk Walked: Reenvisioning Perth's Wetlands." In *Australian Wetland Cultures: Swamps and the Environmental Crisis,* edited by John C. Ryan and Li Chen, 163–174. Lanham, MD: Lexington Books.
Saint Teresa of Avila. 1957. *The Life of Saint Teresa of Avila by Herself.* Translated by J. M. Cohen. London: Penguin. First published in Spanish in 1588.
Sandy, D. B. 2012. "Mountain Imagery." In *Dictionary of the Old Testament: Prophets,* edited by Mark J. Boda and J. Gordon McConville, 554–555. Downers Grove, IL: InterVarsity Press.
Sayers, Andrew. 1998. "Eugene von Guérard 1811-1901: 'Tower Hill 1855.'" In *New Worlds from Old: 19th Century Australian and American Landscapes,* edited by Elizabeth Johns, Andrew Sayers and Elizabeth Kornhauser, 154. Canberra: National Gallery of Australia/Hartford, CT: Wadsworth Atheneum.
Sayers, Dorothy. 1934. *The Nine Tailors.* New York: Harcourt.

Schaffer, Kay. 1988. *Women and the Bush: Australian National Identity and Representations of the Feminine.* Cambridge: Cambridge University Press.

Schenker, Heath. 2009. *Melodramatic Landscapes: Urban Parks in the Nineteenth Century.* Charlottesville, VA: University of Virginia Press.

Scott, James C. 2017. *Against the Grain: A Deep History of the Earliest States.* New Haven, CT: Yale University Press.

Scrase, David. 1979. *Drawings & Watercolours by Peter de Wint: A Loan Exhibition Inaugurated at the Fitzwilliam Museum, Cambridge.* Cambridge: Cambridge University Press.

Semeniuk, Vic. 1993. "Geology, Landforms, Soil and Hydrology." In *Mountains of Mystery: A Natural History of the Stirling Range*, edited by Carolyn Thomson, Graham Hall and Gordon Friend, 13–26. Como, Western Australia: Department of Conservation and Land Management.

Shears, Richard. 2008. *Swamp: Who murdered Margaret Clement?* Chatswood, NSW: New Holland.

Shelley, Percy Bysshe. 1975. *The Complete Poetical Works, Volume II, 1814-1817.* Edited by Neville Rogers. Oxford: Clarendon Press.

Sir Gawain and the Green Knight 1975. Translated by J. R. R. Tolkien. London: George Allen and Unwin.

Solnit, Rebecca. 2005. *A Field Guide to Getting Lost.* New York: Viking.

Sornig, David. 2018. *Blue Lake: Finding Dudley Flats and West Melbourne Swamp.* Melbourne: Scribe.

Standing, Guy. 2019. *The Plunder of the Commons: A Manifesto for Sharing Public Wealth.* London: Penguin.

State Library of New South Wales. n.d.a "Old Tank Stream Sydney 1852?" by John Black Henderson. Accessed April 6, 2020. http://digital.sl.nsw.gov.au/delivery/DeliveryManagerServlet?embedded=true&toolbar=false&dps_pid=IE3238530&_ga=2.179782242.71427905.1600237758-16817892.1600237758.

State Library of New South Wales. n.d.b "Tank Stream, Sydney" by John Skinner Prout. Accessed April 6, 2020. https://www.artgallery.nsw.gov.au/collection/works/1034/.

State Library of South Australia. n.d. "An Artist's Impression of the Tank Stream." Accessed April 6, 2020. https://collections.slsa.sa.gov.au/resource/PRG+280/1/1/205.

Statham, Pamela. 1989. "Patterns and Perspectives." In *The Origins of Australia's Capital Cities,* edited by Pamela Statham, 1–36. Cambridge: Cambridge University Press.

Stewart, Dean. (2020). "The Once as It Was: The Ancestral Family Estates of Melbourne's First Peoples." In *Rod Giblett Modern Melbourne: People and Places of Nature and Culture,* map inside front cover. Bristol: Intellect Books.

Sussex, Lucy. 2015. *Blockbuster! Fergus Hume and the Mystery of a Hansom Cab.* Melbourne: Text Publishing.

Swift, Graham. 2010a. "Introduction to the 25th Anniversary Edition." In *Waterland: With a New Introduction by the Author,* v–xi. London: Picador.

Swift, Graham. 2010b. *Waterland: With a New Introduction by the Author*. London: Picador.

Tearle, Nick. n.d. "Fenland Paintings." Accessed January 7, 2020. https://nicktearle.net/fenland-paintings.

Thayer, Harvey W. 1954 "The Blue Flower." *The Encyclopedia Americana*. Accessed February 29, 2020. http://www.roangelo.net/logwitt/waterman.html.

Thompson, E. P. 1968. *The Making of the English Working Class*. Harmondsworth: Penguin.

Thompson, Ian H. 2014. *Landscape Architecture: A Very Short Introduction*. Oxford: Oxford University Press.

Thoreau, Henry David. 1958. *The Correspondence,* edited by Walter Harding and Carl Bode. New York: New York University Press.

Thoreau, Henry David. 1962. *The Journal 1837-1861, Volumes I-XIV,* edited by Bradford Torrey and Francis Allen. New York: Dover.

Thoreau, Henry David. 1980. *Natural History Essays*. Salt Lake City: Gibbs Smith.

Thoreau, Henry David. 1982. *The Portable Thoreau,* edited by Carl Bode. New York: Viking Penguin.

Thoreau, Henry David. 1993. *Faith in a Seed*, edited by Bradley Dean. Washington, DC: Island Press/Covelo, CA: Shearwater Books.

Thoreau, Henry David. 1997. *Walden*. Boston: Beacon. First published in 1854.

Thoreau, Henry David. 2010. *Collected Poems*. Edited by Carl Bode, enlarged edition. Baltimore: Johns Hopkins University Press.

Thoreau, Henry David. 2013. *The Correspondence: Volume 1: 1834-1848,* edited by Robert Hudspeth. Princeton: Princeton University Press.

Thorson, Robert. 2018. *The Guide to Walden Pond*. Boston: Houghton Mifflin.

Tillyard, Sylvia. 2018. *The Great Level*. London: Chatto & Windus.

Tolkien Gateway, The. n.d. "Dead Marshes." Accessed September 11, 2014. http://tolkiengateway.net/wiki/Dead_Marshes.

Tolkien, J. R. R. 1977. *The Silmarillion*. London: George Allen and Unwin.

Tolkien, J. R. R. 1981. *Letters of J.R.R. Tolkien: A Selection*. Edited by Humphrey Carpenter with the assistance of Christopher Tolkien. London: George Allen and Unwin.

Tolkien, J. R. R. 2002. *The Two Towers: Being the Second Part of the Lord of the Rings*. London: HarperCollins. First published in 1954.

Tolkien, J. R. R. 2006. "*Beowulf*: The Monsters and the Critics." In *The Monsters and the Critics and Other Essays,* edited by Christopher Tolkien, 5–48. London: HarperCollins. First published in 1936.

Tolkien, J. R. R. 2014. *Beowulf: A Translation and Commentary*. Edited by Christopher Tolkien. London: HarperCollins.

Turnbull, Lucy Hughes. 1999. *Sydney: Biography of a City*. Milsons Point, NSW: Random House Australia.

Underwood, Roger and Andrew Burbidge. 1993. "Introduction." In *Mountains of Mystery: A Natural History of the Stirling Range*, edited by Carolyn Thomson, Graham Hall and Gordon Friend, 3–4. Como, Western Australia: Department of Conservation and Land Management.

Walden Woods Project, The. 1991. *Walden Woods: Cradle of American Conservation*. Boston: The Walden Woods Project.

Walls, Laura Dassow. 2017. *Henry David Thoreau: A Life*. Chicago: University of Chicago Press.

Waterman, Tim. 2015. *The Fundamentals of Landscape Architecture*, second edition. London: Bloomsbury.

Webb, Hugh 1996. "Aboriginal Country: Not a Construction, A Way of Being." In *Western Australian Wetlands: The Kimberley and South-West*, edited by Rod Giblett and Hugh Webb, 61–76. Perth, Western Australia: Black Swan Press/ Wetlands Conservation Society.

Western Australian Museum. 2014. "Reimagining Perth's Lost Wetlands." Accessed May 1, 2015. http://museum.wa.gov.au/explore/wetlands.

Wettenhall, Gib. 1999. *The People of Gariwerd: The Grampians' Aboriginal Heritage*. Melbourne: Aboriginal Affairs Victoria.

Wilkie, Benjamin. 2020. *The Grampians: An Environmental History*. Clayton South, Victoria: CSIRO.

Williams, Merryn and Raymond Williams, editors. 1986. *John Clare: Selected Poetry and Prose*. London: Methuen.

Williams, Raymond. 1973. *The Country and the City*. London: Chatto & Windus.

Woolnough, W. G. 1920. "Geological Reconnaissance of the Stirling Range of Western Australia." *Journal and Proceedings of the Royal Society of N[ew] S[outh] Wales* no. 54: 79–113.

Wraight and Associates. n.d. "Jellicoe Street, North Wharf & Silo Park." Accessed November 26, 2017. http://www.waal.co.nz/wsud/jellicoe-street-north-wharf-silo-park/.

Yale Center for British Art. n.d. "Cotman." Accessed January 7, 2020. https://collections.britishart.yale.edu/?utf8=✓&join=AND&lookfor0%5B%5D=&type0%5B%5D=allfields&bool0%5B%5D=AND&lookfor1%5B%5D=&type1%5B%5D=title&bool1%5B%5D=AND&lookfor2%5B%5D=Cotman&type2%5B%5D=auth_&bool2%5B%5D=AND&lookfor3%5B%5D=&type3%5B%5D=earliestDate&bool3%5B%5D=AND&lookfor4%5B%5D=&type4%5B%5D=topic&bool4%5B%5D=AND&lookfor5%5B%5D=&type5%5B%5D=geographic&bool5%5B%5D=AND&search_field=all_fields&q=Cotman.

Index

About the Author

Rod Giblett is the author of *Postmodern Wetlands: Culture, History, Ecology* (Edinburgh University Press, 1996) in which he critiqued the mainstream denigration of wetlands in western culture and presented the counter-tradition in it and Australian Aboriginal cultures. Now regarded as a classic in the environmental humanities, it was the first book of wetland cultural studies and psychoanalytic ecology. Ranging across philosophy, history, literature, and psychoanalysis, and engaging with French feminism and post-structuralism, it considered wetlands in relation to aesthetics, the city, the body, the mind, war, politics, and nature writing in England, America, and Australia.

Rod is also the contributor of chapters on wetlands in Australian Aboriginal cultures and Australian landscape painting and photography to the edited collection *Australian Wetland Cultures* (Lexington Books, 2020). *Wetlands and Western Cultures* develops and expands Rod's previous work into new geographical areas and topics. It is the celebration of twenty-five years since the publication of *Postmodern Wetlands* and the culmination of thirty years of research and writing about wetland cultures. Rod Giblett is honorary associate professor of Environmental Humanities in the Writing and Literature Program of the School of Communication and Creative Arts at Deakin University, Victoria, Australia.

www.ingramcontent.com/pod-product-compliance
Ingram Content Group UK Ltd.
Pitfield, Milton Keynes, MK11 3LW, UK
UKHW020840290926
15470UKWH00004B/1470

9 781793 643452